Naturalists' Handbooks 27

Insects on cherry trees

SIMON R. LEATHER
Department of Biology,
Imperial College,
Silwood Park,
Ascot SL5 7PY

and

KEITH P. BLAND
35 Charterhall Road,
Edinburgh EH9 3HS

with illustrations by Miranda Gray

T0145993

Published for the Company of Biologists Ltd by

The Richmond Publishing Co. Ltd

P.O. Box 963, Slough SL2 3RS

Series editors
S. A. Corbet and R. H. L. Disney

Published by The Richmond Publishing Co. Ltd,
P.O. Box 963, Slough SL2 3RS, ENGLAND
Telephone: 01753 643104
email: rpc@richmond.co.uk

ISBN 0 85546 311 2 Paper
ISBN 0 85546 312 0 Hardcovers

Printed in Great Britain

Contents

Editors' preface

Cherry trees of various kinds are common in gardens and in the countryside, and the rich communities of insects that live on them offer excellent opportunities for research. The authors' own work has already shown how studies of cherry insects can address major ecological questions. As cherries become more popular for planting in amenity areas and woodland, it becomes increasingly important to understand how their insects interact with each other and with the host plant. We hope this book will encourage more people to contribute to this research and to take advantage of the opportunities offered by this very accessible and diverse insect community. The book introduces the insect species associated with cherry trees, draws attention to particularly interesting features of the insect communities, and highlights topics on which further investigation is needed.

S.A.C.
R.H.L.D.
July 1998

Acknowledgements

It is a pleasure to thank Sally Corbet and Henry Disney for their invaluable help with the preparation of this book. It would have been impossible without them. David Huggett helped with the graphics, in particular the leaves and flowers. His expertise with the vagaries of the Silwood scanner are much appreciated. Finally, but not least – many thanks to our families for putting up with our eccentric interests in the fauna of cherry trees.

S.R.L.
K.P.B.

1 Introduction

Several different species of cherry are grown in Britain. The great majority of these are grown as ornamental trees and shrubs because of the fine display their flowers make in the early spring (pl. 1), and their suitability for small gardens. Although some species of cherry bear delicious fruit, nowadays few trees are grown for fruit outside commercial orchards. This was not always the case: fruit-bearing cherry trees were a familiar feature of Victorian gardens. Wild cherries are common in our hedgerows and roadsides where they provide attractive blossom in the spring, and a valuable source of food for birds later in the year. In fact, the presence of so many cherries in hedgerows is a direct result of their popularity with birds as a food supply, as after passing through a bird's stomach the cherry stone, which remains relatively unscathed, is often deposited in an area very suitable for germination the next spring.

History

Many of the cherries planted in gardens are introduced species of Asian origin or are cultivated varieties of the native British species (Mitchell, 1974)*. Most garden cherries have been bred or selected for their flowers, mainly for an increase in number or size and for pink and red coloration.

Worldwide there are over 200 species of cherry, but only six species grow naturally in the British countryside: *Prunus avium* the wild cherry, or gean as it is called in Scotland, *Prunus cerasus* the sour cherry, *Prunus cerasifera* the cherry plum or Myrobalan plum, *Prunus laurocerasus* the cherry laurel, *Prunus padus* the bird cherry and *Prunus serotina* the black cherry.

Prunus avium and *P. padus* are true natives of Britain, fossil records of their pollen and seeds indicating that they have been present since the end of the last ice age (Godwin, 1975). *Prunus cerasus* was introduced by the Romans and *P. laurocerasus*, *P. cerasifera* and *P. serotina* were introduced from Europe in the 15th, 16th and 17th centuries respectively as the popularity of ornamental gardens grew.

Prunus avium and *P. padus* are very common trees in both their wild and domesticated forms and are found throughout Britain (Leather, 1992). *Prunus padus* is more common in the north of England and in Scotland, although there is a local population in East Anglia. Any bird cherries found south of the Midlands but not in East Anglia are likely to have been planted by humans as there appears to be a natural barrier to their distribution (fig. 1). *Prunus avium*, on the other hand, is native throughout Britain (fig. 2). Between them, these two species occupy almost every 10 km square in the British Isles.

* References cited under authors' names in the text appear in full in Further reading on p. 75.

2

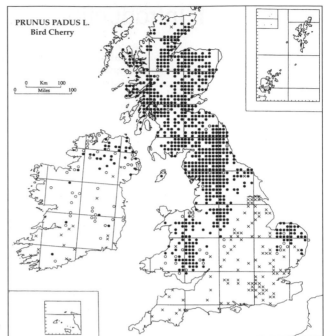

Fig. 1. The distribution of *Prunus padus* in the British Isles. Each dot represents at least one record in a 10-km square of the National Grid. ○: before 1950; ●: 1950 onwards; x: introduced. Mapped by the Biological Records Centre, Institute of Terrestrial Ecology, and reproduced with permission from Blackwell Science. (After Leather, 1996.)

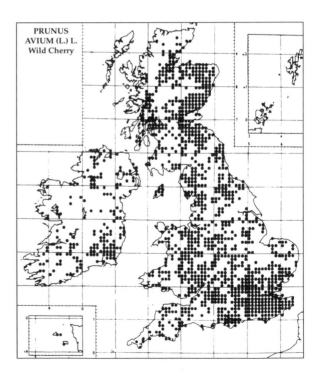

Fig. 2. The distribution of *Prunus avium* in the British Isles. Each dot represents at least one record in a 10-km square of the National Grid. Reproduced with permission from the Botanical Society of the British Isles.

The fruits of the bird cherry, *P. padus*, are harsh and bitter tasting and unpalatable to humans, but are favoured by birds, hence the common English name. Yet the Latin name *P. avium*, surely derived from the word *avis* meaning bird, is associated not with the bird cherry but with the wild cherry, the fruits of which are pleasantly flavoured and a great favourite with humans.

Prunus avium (L.) L. Gean, Wild cherry, Mazzard

Prunus avium is a deciduous tree ranging from 5 to 25 m in height. It is especially common in areas over clay or chalk. It is a very common garden and park tree and is widely distributed through Europe, North Africa and West Asia. The bark is smooth and peels off in thin strips. It is reddish brown, ranging from reddish pink in young trees to dark purple in very old trees. The branches spread or ascend giving young trees a very regular appearance. The shoots are pale, red-brown above and grey beneath, with pointed shiny red brown buds. The leaf stalk is grooved and has two large reddish glands near the tip and the leaves are 6–15 cm long, a light dull green, hairless above but with short hairs below. The flowers, 8–15 mm across, are white and cup-shaped (pl. 1.2). In young trees they are arranged along the shoots in clusters, but in older trees they are more bunched and have longer stalks. They open in early spring, just before the leaves. The flowers are visited by various insects, some of which may act as pollinators. The fruits, usually three to five per bunch, are bright or dark red turning blackish red by the end of summer.

Prunus cerasifera Ehrh. Cherry plum, Myrobalan plum

Prunus cerasifera is normally the first *Prunus* to flower in spring. It is a deciduous shrub or small tree growing to about 8 m. It was introduced to Britain in the 16th century, and can be seen locally in southern England (fig. 3) as a tall hedge or roadside thicket but is less common in gardens. It occurs throughout Russian Central Asia, Iran, and to the Balkans. The shoots are green and the leaves are toothed with blunt or pointed teeth. The flowers appear at the same time as the leaves and are white with a short, purplish green grooved stalk. The fruits, in bunches of one to five, are glossy pale green in summer, turning yellow or reddish by the end of the year. They are produced only in favourable years. Cultivated forms 'Atropurpurea' (purple leaves, pale pink to white flowers) and 'Nigra' (purple leaves, rosy pink flowers) are commonly grown in gardens. Although the flowers are visited by a number of insects it is not certain whether any of these act as pollinators.

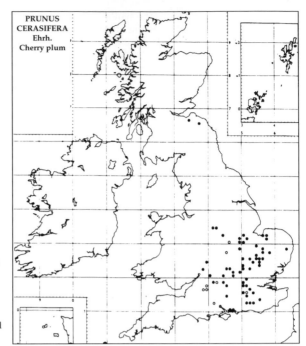

Fig. 3. The distribution of *Prunus cerasifera* in the British Isles. Each dot represents at least one record in a 10-km square of the National Grid. ○ : before 1930; ● : 1930 onwards. Reproduced with permission from the Botanical Society of the British Isles.

Prunus cerasus L. Sour cherry

Sour cherry was introduced to Britain long ago and is usually a shrubby tree, but can reach 7 m in height. It is rather bushy with somewhat drooping branches, and it suckers very easily. It is widespread in hedges in southern England, Wales and Ireland but rarer in northern England and Scotland (fig. 4). It occurs in SE Europe and SW Asia and is regarded as one of the parents of the Morello cherry. The leaves are dark green, 5–8 cm long, and relatively broader than those of *P. avium*. The leaf stalk is short and the glands are small or may even be absent. Sour cherry flowers in April and May, being one of the later cherries to flower. The fruit is bright red and acid, hence the name sour cherry. The flowers are visited by bees, which are the most likely pollination agents.

Prunus incisa Thunb. ex Murray Fuji cherry

This is a large shrub or small tree, 6–10 m high. The leaves are ovate, 4–8 cm long and slightly toothed. The flowers are pink and the fruit is purplish black, ovoid and 6–8 mm in diameter. This species was introduced from Japan. It is naturalised in oakwoods on clay-with-flints soil types. It is much planted in parks and gardens; some very fine examples can be found in Windsor Great Park.

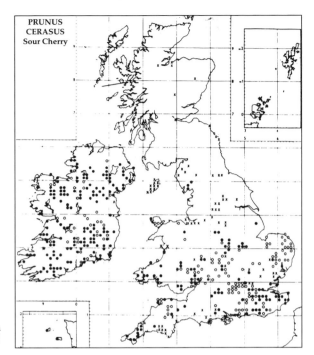

Fig. 4. The distribution of *Prunus cerasus* in the British Isles. Each dot represents at least one record in a 10-km square of the National Grid. ○: before 1930; ●: 1930 onwards; x: introduced. Reproduced with permission from the Botanical Society of the British Isles.

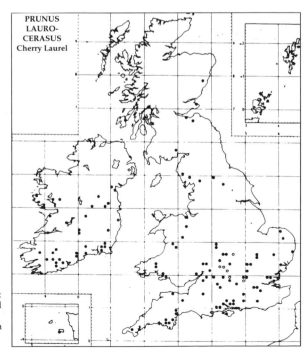

Fig. 5. The distribution of *Prunus laurocerasus* in the British Isles. Each dot represents at least one record in a 10-km square of the National Grid. ○: before 1930; ●: 1930 onwards. Reproduced with permission from the Botanical Society of the British Isles.

Prunus laurocerasus L. Cherry laurel

This was introduced in 1576 from SE Europe and Asia Minor. It is an evergreen shrub or small tree, usually 2–6 m high, but occasionally it forms a tree up to 14 m high. It is closely related to the Portugal laurel, *P. lusitanica*. It is locally abundant (fig. 5) as a shrub or hedge and is sometimes naturalised as an undershrub in woods. The bark is brownish grey with pores (lenticels), the shoots are pale green and the leaves are large (20 cm long and 6 cm wide) and glossy greenish yellow. It has fragrant whitish flowers in erect spikes 5–12 cm long. The flower spikes may begin to emerge in mild weather as early as January, but the flowers do not open until early April or even later. The berries are large, purple-black, and 2 cm in diameter. Cherry laurel's main entomological claim to fame is the fact that the leaves give off copious amounts of hydrogen cyanide gas when crushed or cut, and if placed in a jam jar with a lid, they can make a very effective killing jar.

Prunus mahaleb L. St Lucie cherry

This is a shrub, or rarely a small tree, sometimes reaching 10 m. The leaves are 4–7 cm long, broadly oval, and slightly hairy underneath. They have large conspicuous marginal glands. The white, fragrant flowers occur in spikes (racemes) at the end of short, leafy side shoots. The petals are 5–8 mm long. The fruit is oval, black, 8–10 mm in diameter and bitter-tasting. This species is often found along railway lines.

Prunus padus L. Bird cherry

Bird cherry is a deciduous tree native north of the Midlands, in East Anglia and in Ireland. It is common by streams in limestone areas such as NW Yorkshire and the Peak District. It is planted frequently in gardens and sometimes in streets, and in recent years has become common alongside motorways and trunk roads. It is native to Europe and Asia Minor, extending further north than any other *Prunus* species. The bark has a characteristic strong foetid smell; once smelt this is never forgotten, although it is not particularly unpleasant. The bark is a smooth dark grey-brown, and on young twigs it is frequently heavily pockmarked with lenticels. The shoots are shiny and dark brown with sharp compact buds that normally lie close to the stem. They are dark at the base but paler at the tip. The leaves are about 10 cm long and 7 cm wide, with sunken veins and a stout dark red grooved stalk. There are two prominent red-brown glands (extra-floral nectaries) at the base of the leaf or top of the leaf stalk. The fragrant white flowers grow in dense clusters on drooping spikes 8–15 cm long, usually with a few leaves at the base (pl. 1.1). They usually open in late May. The globular black astringent-tasting fruit is a favourite with several bird species, especially thrushes. A large number of varieties are planted, the most common being 'Watereri' which has very long flower spikes and is much prized when in flower.

Prunus pensylvanica L. Pin cherry

This is a deciduous tree sometimes reaching 12 m in height. The leaves are relatively large, 6–12 cm long, ovate and rounded at the base. The flowers are white and are borne on stalks up to 15 mm long, in clusters or sometimes in very short racemes. The fruit is about 6 mm in diameter and red. This species was introduced from North America and is naturalised in woodlands in Surrey, but elsewhere it is found only in parks and gardens.

Prunus serotina Ehrh. Black cherry, Rum cherry

Black cherry is native to NE America and was introduced to Britain in 1629. It is not very common, being largely confined to large or botanic gardens. It is planted in some places for amenity or as the shrub layer for game coverts. It seeds itself and so spreads. It is a deciduous tree up to 30 m high. Early on in the life of the tree the bark is dark purplish grey and peels; in older trees it is a widely fissured brownish pink. The bark is bitter and aromatic but, like that of the bird cherry, not unpleasant to smell. The shoots are slender, purple-brown above and green below. The bud is conical, green and brown and the leaf, 12 cm long with incurved teeth, is a dark glossy green above and paler underneath. The midrib is prominent with bright orange or white hairs in patches along each side of the lower surface of the leaf. The leaf stalk varies in length from being almost absent to 1.5 cm, with glands at the leaf base enlarged into several green wings or minute lobes. The small white flowers are on an erect spike 10–15 cm long. The fruits are globular, 1 cm in diameter and a glossy red. Some of them become dark purple later in the season.

Prunus serrulata Franch. Tibetan cherry

The so-called Tibetan cherry is actually a native of Western China. It was introduced into Britain in 1908 and 1913. It is relatively infrequent, being mainly confined to gardens and some parks. It is a deciduous tree ranging from 10 to 15 m in height. This tree is characterised by the narrowness and fine toothing of the leaves which are lance-shaped, 4–12 cm long but only 1–3 cm wide, and rounded or wedge-shaped at the base. The leaves are downy along the midrib beneath and in the chief vein axils. There are several large glands at the base near the stalk, which is 0.5 to 0.75 cm long. The flowers are white, about 1.5 cm across, and usually produced in twos or threes (sometimes singly or in fours) during April, each on a stalk 1 to 1.5 cm long. The fruits are

Table 1. *Insects recorded feeding on cherry species in Britain.* Further details of nomenclature are given in the checklist on p. 72.

	avium	cerasifera	cerasus	laurocerasus	padus

Hemiptera (bugs, including aphids and allied insects)

	avium	cerasifera	cerasus	laurocerasus	padus
Aguriahana stellulata	+				
Alebra wahlbergi	+				
Alnetoidia alneti	+			+	
Brachycaudus cardui	+				
B. helichrysi		+	+		
Edwardsiana prunicola	+				
E. rosae	+				
Globiceps flavomaculatus		+			
Hyalopterus pruni	+		+		
Lygocoris pabulinus	+				
Myzus cerasi	+		+		
M. padellus					+
M. persicae		+			
Orthotylus marginalis	+				
Pentatoma rufipes	+				+
Phorodon humuli			+		
Phytocoris dimidiatus			+		
Rhopalosiphum insertum			+		
R. padi					+
Typhlocyba quercus	+				+
Zygina species	+				

Lepidoptera (butterflies and moths)

	avium	cerasifera	cerasus	laurocerasus	padus
Abraxas grossulariata					+
Acleris rhombana				+	
A. notana			+		
A. umbrana					+
Adoxophyes orana			+		
Allophyes oxyacanthae			+		
Alsophila aescularia	+				
Argyresthia pruniella			+		
Coleophora coracipennella	+				
Cydia funebrana	+		+		
C. prunivorana			+		
Diloba caeruleocephala				+	
Enarmonia formosana	+			+	
Epiblema trimaculana			+		
Epinotia signatana	+				+
Hedya dimidioalba	+				
H. pruniana	+				
Lomographa bimaculata	+				
L. temerata					+
Lyonetia clerkella	+		+		
Nymphalis polychloros	+				
Operophtera brumata					+
Pammene germmana			+		
P. rhediella			+		
Pandemis cerasana	+		+		

	avium	cerasifera	cerasus	laurocerasus	padus
Phtheochroa schreibersiana					+
Phyllonorycter cerasicolella	+				
P. sorbi			+		+
Scythropia crataegella	+				
Stigmella prunetorum	+		+		
Swammerdamia caesiella	+		+		
S. pyrella			+		
Tischeria gaunacella			+		
Triphosa dubitata			+		+
Xestia baja					+
Yponomeuta evonymella					+
Y. padella	+				

Diptera (flies)

	avium	cerasifera	cerasus	laurocerasus	padus
Asphondylia prunorum		+	+		
Dasineura tortrix		+	+		
Phytobia cerasiferae		+			
Putoniella pruni			+		

Hymenoptera (sawflies)

	avium	cerasifera	cerasus	laurocerasus	padus
Caliroa cerasi	+				
Pamphilius sylvaticus					+
Pristiphora retusa					+

Coleoptera (beetles)

	avium	cerasifera	cerasus	laurocerasus	padus
Anthonomus bituberculatus					+
A. humeralis	+		+		+
Furcipus rectirostris					+
Magdalis cerasi	+		+		
M. ruficornis					+
Meligethes coracinus	+		+		
Ramphus pulicarius	+		+		
Rhynchaenus testaceus	+		+		
Rhynchites aequatus	+		+		+
R. auratus	+		+		+
R. pauxillus					+
Saperda scalaris					+
Scolytus rugulosus	+	+			+
Trachys minutus	+				

oval, 1 to 1.5 cm long on a 4 cm stalk, bright green in summer turning red in autumn. Young trees very characteristically possess a beautiful bright brown peeling bark. It can often be mahogany-red or even orange-brown in colour when it is stroked and caressed frequently. The shoots are covered in a fine down when young. This tree is grown solely for the beauty of its bark.

A key to help with the identification of cherry trees is presented in Chapter 4.

Insects associated with British cherries

Associated with cherries are numerous species of insects that feed on the flowers, fruits, leaves, wood or roots, or on other insects. Some of the insects found on cherry trees are characteristic of these plants, being found more often there than anywhere else. Others are tourists that happened to be on cherries when found but which are more often found elsewhere. Table 1 lists some of the insects that commonly feed on British cherry trees. It is by no means a fully comprehensive list, but it does cover the main groups of insects that are likely to occur on cherries. No one cherry tree will have all these insects on it at the same time, or even through the life of the tree. Only a proportion of the listed insect species will feed on any individual cherry tree. This proportion will vary according to the size and location of the tree, the geographic range of the insects and the time of year. These factors will be discussed in greater detail in the following chapters. The number of insects, both species and individuals, found on an individual tree will also depend on the intensity of the search. As the search is extended and more and more time is spent scanning more and more leaves, the number of insect species found will increase until a ceiling is reached beyond which further searching will add no new species to the list. A search of this kind is worth doing and the procedure is considered further in Chapter 5.

2 Island biogeography and species–area studies

Biogeography has been defined as 'the study of the patterns of distribution of organisms in space and time' (Cox, Healey & Moore, 1976). In the early days of natural history, naturalists were perfectly happy to map animal and plant distribution from the results of sampling. This in itself is a very useful exercise and produces maps similar to figs. 1–5. However, distribution maps cannot explain why particular plants and animals are distributed in the patterns that are seen today. To understand this, detailed studies of habitat requirements must be undertaken.

Islands

In 1835 Charles Darwin was leaving the Galapagos Islands. Impressed by the variation among the skins of the birds he had collected from the different islands, he wrote in his notebook *'when I see these Islands in sight of each other, and possessed of but a scanty stock of animals, tenanted by these birds, but slightly differing in structure and filling the same place in Nature, I must suspect they are only varieties...the Zoology of Archipelagos will be well worth examination'*. Many people regard this as the first reference to evolution and to island biogeography, a field in which major advances in evolutionary and ecological theory have arisen.

Islands are convenient units to study. They are comparatively easy to survey and, for organisms that cannot fly or swim from one island to another, population studies are relatively easier on islands than on the mainland. Important and interesting work has been achieved by scientists all over the world who have made islands their particular area of study. Whatever its size an island is a discrete unit separated from the mainland by water or some other barrier. One of the earliest findings to emerge from the study of islands was that the bigger they were, the more species of plants and animals could be found on them. Great Britain has 44 species of native terrestrial mammals (extant or only recently extinct) plus 13 bats. Ireland, just next door, has only 22 mammal species and only seven bat species. The similar picture shown by the bats rules out the possibility that the shortfall in species number in Ireland is due to the difficulties of crossing the 20 mile stretch of the Atlantic that separates it from Great Britain. Birds, which have little difficulty in reaching Ireland and do so frequently, show a similar pattern. Thirty-nine of the species that commonly breed in Britain have never bred in Ireland (Gorman, 1979). So what is the difference between the two islands? The most obvious one is that Great Britain is much larger than Ireland. Throughout the world, larger islands generally have more species than smaller ones.

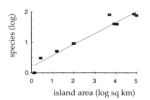

Fig. 6. The relationship between island area (square kilometres) and the number of West Indian species of reptile and amphibian.

Darlington (1957) was able to demonstrate this very clearly for West Indian faunas of reptiles and amphibians (fig. 6). It is interesting to note that Trinidad has many more species than would be expected from its size. This is because Trinidad was joined to South America only 10,000 years ago and has not yet settled into its equilibrium species number. On the basis of these results Darlington developed a neat rule of thumb – for every ten-fold increase in land area, the number of species supported by an island will double. More generally, if the number of species is plotted against island area, using a log scale on both axes, a straight line relationship results. This relationship is described in simple mathematical terms as:

$$S = CA^Z$$

or by taking logs to give the linear form:

$$\log S = \log C + z \log A$$

where S is the number of species, C is a constant giving the number of species when A has a value of one, A is the area of the island and z is the slope of the regression line relating S and A.

It was not until 1961 that Southwood extended the concept to embrace the number of insect species associated with British trees. His simple but important suggestion was that a species of tree was the ecological equivalent of an island to those insects that feed upon it. The barrier of sea that must be crossed by immigrants to a real island is represented by the chemical defences of the tree, which must be overcome by insects colonising from other host plant species. From this simple beginning there has arisen a whole field of biogeography relating to the establishment of nature reserves and conservation areas.

As a measure of the range or abundance of each tree species, equivalent to the area of an island, Southwood took the number of pollen records in Quaternary deposits. The Quaternary era is the most recent period in the geological time scale and thus the pollen fossils found in it are likely to represent modern tree species. Pollen maps have been produced for several plant species in the British Isles (Birks, Deacon & Peglar, 1975; Birks, 1989). Some authors have questioned the validity of extending this measure to represent the abundance of modern tree species (Birks, 1980) but the number of fossil pollen records has been shown to be significantly correlated with the past and present abundance of British tree species (Leather, 1985). Southwood obtained the number of insect species associated with each tree species from published host plant records. His data source has been shown to be very robust (Lawton and others, 1981; Leather, 1990a), so his results may be regarded as reliable. Basically, Southwood showed that the more abundant the tree species was, the more different insect species were associated with it (fig. 7). This concept was further extended by two American

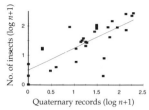

Fig. 7. The relationship between the number of Quaternary pollen records and the number of insect species associated with British tree species. (After Southwood, 1961.)

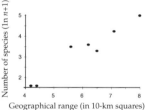

Fig. 8. The relationship between plant range (number of 10-km squares) and the number of insect species associated with British *Prunus* species. (After Leather, 1985, with permission from Blackwell Science.)

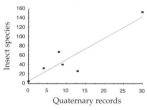

Fig. 9. The relationship between the number of Quaternary pollen records and the number of associated insect species feeding on British cherry trees. (After Leather, 1985, with permission from Blackwell Science.)

scientists (Strong & Levin, 1979) who took advantage of a data source published after Southwood's 1961 study, namely the *Atlas of the British Flora* (Perring & Walters, 1962).

This atlas has proved to be a continuing source of inspiration to entomological biogeographers. Basically the British Isles are divided up into 10-km grid squares and if a particular plant species has been recorded from that grid square a dot is placed on the map (see fig. 1 for an example). There is no indication of how common a particular plant species is in that square – the dot in a particular 10-km square could represent one individual plant or a thousand. Nevertheless, the atlas does give an idea of the range of a particular species and this can be equated to the area covered by a plant species. Strong and Levin used the same insect data sources that Southwood had used in 1961 but calculated the geographic range of the trees from the atlas. They were able to confirm Southwood's observation that the greater the range of a tree species was (the more 10-km squares it occupied), the greater was the number of insect species associated with it. This has also been shown for the different species of cherry trees (fig. 8).

In all the many studies of this nature carried out since then, area or geographic range has been shown to be the most important single factor determining insect species richness on plants. This does not mean that other factors are not important. For example, plant chemistry, the length of time the host plant species has been present in a particular geographic area ('island age') and plant architecture may also influence the number of insect species associated with a particular plant species (Strong, Lawton & Southwood, 1984). *Prunus avium*, which is native to Britain, would be expected to have considerably more species of insects feeding on it than *Prunus laurocerasus* which was not introduced until the 16th century. This is indeed the case (table 1). As already shown by Southwood (1961), the number of insects associated with a particular plant species depends on the history of the plant (island age), and the number of insect species recorded as feeding on British *Prunus* species is highly correlated with the pollen records for the trees (fig. 9). In general, the more widely distributed a plant species, and the longer it has been present in Britain, the greater is the number of insect species that have adapted to feed on it. Perhaps this is why a tree like the cherry laurel, *P. laurocerasus*, which is a recent addition to the British landscape and is not particularly widespread, has only four species of insect recorded as feeding on them. However, in this case there may be another explanation. Cherry laurel leaves contain so much hydrogen cyanide that some amateur entomologists use them in their killing jars, so it is perhaps not surprising to find so few insects feeding on them. The Myrobalan plum, *P. cerasifera*, which is a palatable tree, is a more straightforward example. It is also recorded as having few insect species feeding on it. It is a recent arrival in Britain and not widespread, so it fits the theory very well.

Apparent anomalies such as the case of the cherry laurel forced entomologists to explore the influence that other plant characteristics might have in determining insect species richness. These included plant chemistry (Jones & Lawton, 1991), taxonomic isolation, how many close relatives the plant possessed in the geographic area (Kennedy & Southwood, 1984), and plant architecture, the number and size of leaves and plant height (Lawton, 1983). The species–area relationship is further improved if the analysis is confined to one particular group of plants or insects such as leaf-mining Lepidoptera on American oaks (90% of the variance explained) (Opler, 1974), or all insects on a single plant genus such as *Prunus* (85% of the variance explained) (Leather, 1985), or on a single species, such as bracken (96% of the variance explained) (Lawton, 1984).

Whatever the approach used, plant area or abundance always accounted for more variation in insect species number than any other factor. The amount of variation explained in this way is always significant but sometimes quite low. For example, in the case of agromyzid flies whose leaf-mining larvae feed on British umbellifers, it accounts for only 32% of the variation in insect species numbers (Lawton & Price, 1979). In an attempt to explain more of this variation in the numbers of species of agromyzid flies on British umbellifers, Fowler and Lawton (1982) incorporated the concept of local abundance into their equations. They estimated local abundance by scoring the number of tetrads (2 km x 2 km squares on the map) occupied by each plant species within occupied 10 km squares, deriving the information from local Floras which were available for eight English counties. The maximum number of tetrads that can be so occupied within the 10 km square is 25, and the minimum is one. They defined local abundance (b) as:

$$b = \frac{\text{total number of occupied tetrads}}{\substack{\text{total number of available tetrads} \\ \text{within occupied 10 km square}}} \times 100$$

Thus a plant species found in only a few 10 km squares, but occupying every tetrad within them, has a local abundance of 100%; a species recorded from only one or two tetrads within each 10-km square has a very low local abundance, perhaps less than 10%. They found that widespread British umbellifers with large geographic ranges also tended to be locally abundant and vice versa. Multiplying size of geographic range (number of occupied 10-km squares) by local abundance gives an estimate of the total geographic range of the host plant in tetrads. Substituting this value in the species–area equation provided only a marginally better fit for this study and in others that used the same method, for example insect species on Rosaceae (Leather, 1986).

Similarly, it is possible to improve the relationship between area or range of plant and the number of insect species associated with that plant by taking into account variables such as plant height or leaf size (usually geometric mean leaf size, that is, the antilog of the mean log size). Lawton & Schroder (1977) found that trees have more insect species recorded for them than shrubs, and shrubs have more insect species than herbs. To give a better picture of the leaf biomass available for the insects to feed on, others have tried to combine plant height (*h*) and leaf length (*l*) into one index, plant complexity (*c*), where

$$c = \log (hl^3)$$

In some cases this increases the amount of variance explained (Leather, 1986).

Plant chemistry has been invoked to help to explain why certain plants or plant groups have more or fewer insect species than would be expected by virtue of their geographic range (Southwood, 1961; Leather, 1985). However, in another analysis on British umbellifers, Jones & Lawton (1991) could not confirm that biochemically diverse or unusual British umbellifers supported less species-rich assemblages of insect herbivores. They saw some indication that chemically diverse plants had more insect species but the evidence was very weak.

How do insects find the plants on which to feed or lay their eggs? Usually herbivorous insects are considered to recognise their host plants largely by chemical signals, picked up either before alighting or by actual contact with the leaf or plant surface. Using the Sorensen similarity index (method, p. 69) it is possible to construct a dendrogram based on the insect species recorded as feeding on a particular group of plants. This index measures the similarity in terms of the number of insect species that a pair of trees or a group of species of cherry trees have in common. Once the similarity indices have been calculated it is possible to look at the relatedness of the trees in a number of ways. A dendrogram is probably the easiest way to approach this.

For certain members of the genus *Prunus* (fig. 10) the relationships defined by total insect species usage were found to coincide exactly with the classification of *Prunus* down to the sub-generic level as accepted by botanists. Probably these patterns of usage by insects depend on

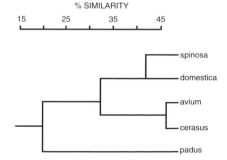

Fig. 10. Dendrogram showing relationships among species of *Prunus* as determined by similarities of insect load. (After Leather, 1985, with permission from Blackwell Science.)

chemical differences which insects can recognize, even if the human observers cannot. These patterns may also be apparent when examining the introduced members of the genus, and perhaps even when examining only one particular insect group or guild.

Often, plant ranges estimated on the basis of presence and absence data do not include man-made plantings. In some cases these could influence the figures dramatically. If these 'real' figures were used we might expect the amount of variance explained by the area, abundance or range of the host plant to be greatly increased. Claridge & Evans (1990) attempted this for British trees by using the Forestry Commission woodland census, which gives figures for the actual area covered by certain tree species and includes natural planting, forest plantations and hedgerow trees. To their surprise the use of this 'improved' data *reduced* the amount of variation in insect species number explained by tree area. This unexpected finding should be examined in future studies.

Species–area work requires little equipment and can be done by simple sampling techniques as described in Chapter 5. For example species–area curves can be constructed in a number of ways. Insects can be collected from selected trees using a predetermined sampling effort, for example a set number of hits with the beating stick or a set number of leaves examined by eye. This can be done for a number of different species of trees and the insects caught can be related to the faunal lists available or the geographic range of the species, if this is shown in the *Atlas of the British Flora*. Alternatively, the data obtained could be related to local abundance, that is, how many trees of that species are in the immediate area. If the same trees are sampled at different levels of intensity it is possible to investigate the extent to which the number of species or individuals of the same species obtained depends on sampling effort. It is also likely that patch size (tree size, or even clump size) may affect the results.

Fractal ecology and the importance of plant architecture

Why should trees have more insect species associated with them than shrubs? Perhaps the most obvious answer is that trees are easier for insects to find than shrubs and herbs because they are bigger. Also, a tree has a larger collecting area for an insect to alight upon. However, this cannot be the whole story. Some species of grasses occur in large, almost single-species stands in meadows in nature as well as in farmland. Surely these are very easily found and present very large collecting areas for insects, yet the number of insect species may not be high. Plant chemistry may be the answer yet again. Van Emden (1990), sampling grass heads for aphids in uncultivated areas, found that most of the grass plants sampled had no aphids present. It seems to be generally true that insects are sparse in natural vegetation.

Herbivorous insects are in general host-specific, living on one or a few closely related species of plants. Even within a species of plant, every genetically distinct individual is different in terms of chemical and architectural characteristics, and thus more or less suitable for a particular insect species or individual. The landscape as seen by an insect is a vegetational map with few host plants present and these vary in quality. The distance between the plants is often large in relation to the size of the insect. Thus host plants may be hard to find, and large plants are more likely to be found than small plants, especially by small insects.

In a simple shape such as a square or a circle, there is no increase in the amount of detail that can be seen when the resolution or magnification is increased. Real objects are usually more complex, even if they seem uniform at a casual glance. For example, a plain painted wall seen from several metres away appears to be a flat surface, but at closer range, the brush or roller strokes become visible. Closer still and the paint 'speckles' that make up the roller and brush strokes become visible and so do cracks and crevices. Under a magnifying glass a landscape of hills and valleys begins to appear, and if a microscope is used the complexity of the environment increases still further. A structure is described as 'fractal' if its length or area becomes disproportionately larger as the unit of measurement is decreased. If a figure with fractal geometry such as a map of the British Isles is measured the answer obtained will depend on the precision with which the length of the coastline is determined. If a ruler marked in centimetres is used the answer obtained is very different from that obtained using a ruler marked in millimetres. If the map is outlined with a piece of string which is then measured, the answer is larger still, and if a piece of finer thread is used an even greater distance will be recorded. The length of the outline increases with the resolution of its detailed convolutions.

What has this to do with biogeography, species–area curves and plants and insects? There are two closely linked implications. Firstly, fractal geometry may enable niche structure to be quantified; and secondly, it may allow us to estimate the limit to the size of animals that can feed or live on a particular plant or island. Fractal dimensions of plants can be calculated in a number of ways. Perhaps one of the simplest is that devised by Morse and others (1985). They photographed plants in such a way that they appeared two dimensional, and then placed the photograph under a grid. The number of squares entered by the outline of the silhouette of the plant were then counted, first starting with the biggest square, then the next size down and so on (fig. 11). An approximate fractal dimension can then be calculated. To a squirrel an oak tree offers a number of places in which to live and forage but a single branch is of little use except in transit. A beetle might spend its whole life on a branch, but a single leaf would support it for only a few days. An even smaller creature, such as an aphid, could spend its whole life on the leaf. And so on.

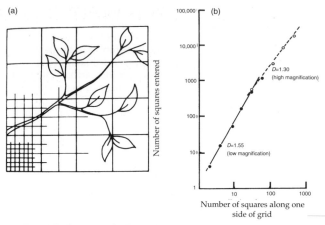

Fig. 11. To determine the fractal dimension of a plant, place a grid over a photograph of the plant (a). Count the number of squares entered by the outline of the plant, starting with a coarse grid of two large squares on one side; then move to finer grids of four, eight or 16 squares. The log number of squares entered by the outline can then be plotted against the log number of squares along one side of the grid (b). The slope of the line is D, the fractal dimension.

Mathematically, D, the fractal dimension of the plant, can be related to animal body length, L, and the perceived area of the habitat, p, like this:

$$p = 1/L^{(1-D)}$$

Thus for a fractal surface with a fractal dimension (D) of 1.5 times the area perceived by an insect 3 mm long may be up to ten times greater than that perceived by an animal 30 mm long. It is possible to predict how many animals could live on the vegetation, if one makes assumptions about how metabolic rate changes with body size. By doing the appropriate sum it is possible to calculate that a ten-fold decrease in animal body size would result in a 178-fold increase in the number of individuals that could feed on a particular resource. Thus, architectural complexity, as well as plant size, is of great importance in determining the number of insects that a particular plant species can support. Trees, being in general more complex than other plants, offer a greater variety of habitats to animals and the possibility of greater community development. This is dealt with in the next chapter.

3 Insect–plant interactions

One of the most important current areas of ecology is the study of the relationship between herbivorous animals and the plants they eat. Some of the most fascinating studies in this area have involved the interaction between insects and plants, and the part these interactions play in shaping plant and animal communities. Not all cherry species are equally suitable for every insect species. Some insect species are highly adapted to a particular host plant and are unable to survive on anything else; others have a wider host range. However, it is unlikely that even those insect species that have a wide host range do equally well on all the cherry species. The suitability and palatability of cherry species to insect herbivores is an area in which much research is required.

Communities and resource partitioning

As we have seen, in general trees have more species of insects feeding on them than shrubs, and in turn shrubs have more species of insects feeding on them than herbs (Lawton & Schroder, 1977). This is so whether we compare species within a plant family or between them (Leather, 1986). Within the family Rosaceae, trees have more insect species recorded as feeding on them than shrubs which in turn have more insects recorded on them than do herbs. This is true within other plant families too. For example Moran (1980) related the number of herbivorous insect species known to attack species of prickly pear cactus (*Opuntia*) in America to the 'architectural rating' of the plant. Tall species of *Opuntia* with numerous large green leaf-like lobes (cladodes) borne on large woody stems hosted up to 30 species of herbivorous insects, whereas small, architecturally simple species hosted fewer than five species (fig. 12).

Even on a large prickly pear, we would not expect all 30 insect species to coexist on the same plant, because they will not appear at the same time of year or have the same geographical range. What factors determine which species can coexist as members of the community on a given plant? An ecological community has been defined as a group of species living closely enough together to have the potential of interaction (Strong and others, 1984). This key characteristic of a community, the potential for interaction among species, is manifested in competition for resources such as egg-laying sites, overwintering sites or, perhaps most importantly, feeding sites.

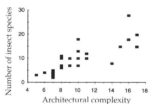

Fig. 12. The number of plant-feeding insect species associated with 28 species of cactus in North and South America, as a function of the architectural complexity of each species of cactus. (Reproduced with permission from Blackwell Science from Strong and others, 1984.)

Niches

A concept that many ecologists find helpful when thinking about communities is that of the niche. What is a niche? *The Oxford English Dictionary* defines a niche variously as a shallow, ornamental recess or hollow in a wall to contain a statue or other decorative object; a small vaulted recess or chamber made in the thickness of a wall or in the ground; a place or position adapted to the character or suited to the merits of a person or thing; or a place of retreat or retirement. It was probably the last two phrases that were in Hutchinson's mind when in 1944 he defined the 'ecological niche' as 'the sum of all the environmental factors acting on an organism', or 'a region of n-dimensional hyper-space, comparable to the phase space of statistical mechanics'. Less challenging and more easily understood is the definition of an ecological niche given in Henderson's *Dictionary of Biological Terms* (1979): the status of an organism in the ecosystem, including its habitat and its effect on other organisms and on the environment.

Competition and resource partitioning

Competition is a 'manifestation of the struggle for existence in which two or more organisms of the same or different species exert a disadvantageous influence upon each other because their more or less active demands exceed the immediate supply of their common resources' (Bakker, 1961). This definition draws attention to the distinction between competition between individuals of the same species, known as intra-specific competition, and competition between individuals of different species, or inter-specific competition. Competition between species for resources which are in limited supply has long been regarded as one of the most important processes determining how many species can coexist on a plant and which feeding niches they occupy, but Strong and others (1984) have recently argued that inter-specific competition is not a major force structuring plant-feeding insect communities. They point out that although green plants are not universally edible, every plant species has some potentially damaging insect species associated with it. The fact that these herbivores are generally scarce for most of the time suggests that their numbers are limited, not by the availability of plants as food, but rather by their natural enemies, predators and parasites. We accept this premise but we suggest that the non-competitive situation described by Strong and others may be the consequence of competition earlier in evolution, resulting in progressive divergence of niches over evolutionary time so that competition no longer exists. We invoke the 'ghost of competition past' (Connell, 1980) to explain the absence of competition now. Furthermore, we suggest that both inter- and intra-specific competition are still alive and operate today to a greater extent than many people think, and that some of the insects feeding on cherries provide evidence to support this claim.

Intra-specific competition

Competition between individuals of the same species, intra-specific competition, has been demonstrated in a number of species. Perhaps some of the most dramatic examples are those to be found in the gall aphids. The aphid *Pemphigus betae* makes galls on the leaves of a type of poplar tree, the narrowleaf cottonwood *Populus angustifolia* in America. The success of a gall depends on its position on the leaf; the closer the gall is to the base of the leaf the larger it will grow, the better are its chances of survival and the more new aphids it will produce (Whitham, 1978). So important is the position of the gall that the founder aphids responsible for initiating the galls actually fight each other for possession of the gall in the best location (Whitham, 1986).

Among the aphids on British cherry trees intra-specific competition is less dramatic, but it does occur in at least one stage of the aphid life cycle. The eggs of the bird cherry aphid *Rhopalosiphum padi* are laid in the angle between the bud and the current year's shoot. Hence they are securely anchored, surrounded on three sides by the plant and protected to some extent from natural enemies and the effects of wind and rain. An average bud can harbour only about 10 eggs in this way. Any extra eggs must then be laid in an exposed position – on the bud, on the bark or on top of another egg. The egg mortality occurring in the first two weeks after leaf fall is significantly correlated with initial egg density, particularly in those years when there are more than 10 eggs per bud (fig. 13). There is thus competition for the best egg-laying sites, in that egg-laying by one individual exerts a disadvantageous effect on another because of the limited availability of good sites. Other examples of intra-specific competition may await discovery. Careful studies of the effects of population density on growth rate, survival or fecundity (the numbers of eggs or young produced per female, or in aphids, ovariole number, p. 68), in the laboratory or in sleeves or clip cages in the field (techniques, p. 67), could show whether or not competition occurs in other species. These studies could provide a basis for evaluating the role of competition more generally among plant-feeding insects.

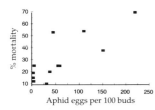

Fig. 13. Relationship between initial egg density of the aphid *Rhopalosiphum padi* and mortality in the first two weeks of winter. (After Leather, 1990b, with permission from Intercept Press.)

Inter-specific competition

To consider inter-specific competition, that occurring between individuals of two or more different species, it is useful to return to the concept of resource partitioning. Insect species can share out or partition a resource, such as a cherry tree, amongst themselves in a number of ways. They can partition it with respect to time or space. For example, of the 26 insect species recorded as feeding on bird cherry (table 1), seven are Lepidoptera that feed only on the leaves, *Abraxas grossulariata, Acleris umbrana, Lomographa temerata, Operophtera brumata, Triphosa dubitata, Xestia baja* and *Yponomeuta evonymella*. We might expect that caterpillars of

Table 2. *Insect species feeding on* Prunus padus: *their specificity, feeding seasons and feeding sites.*
Roman numerals refer to months of the year in which the insect feeds on the plant.
G = generalists, insects that feed also on plants outside the Rosaceae.
R = insects that feed only on plants within the Rosaceae.
* = insects that feed only on *Prunus padus* (*R. padi*, although feeding on grasses in the summer, is host-specific to *P. padus* at all other times of the year).

	Host range	Feeding season	Feeding site
Hemiptera			
Alnetoidia alneti	G	VII–X	leaf mesophyll
Myzus padellus	*	III–VI, VIII–XI	phloem (galls)
Rhopalosiphum padi	*	III–VI, VIII–XI	phloem (galls)
Typhlocyba quercus	G	VII–X	leaf mesophyll
Lepidoptera			
Abraxas grossulariata	G	VIII–V	leaves
Acleris umbrana	G	VI–VII	leaves
Epinotia signatana	R	V	leaves and shoots
Lomographa temerata	G	VII–VIII	leaves
Operophtera brumata	G	V–VI	leaves and buds
Phtheochroa schreibersiana	G	VI–IX	leaves and shoots
Phyllonorycter sorbi	R	VII–X	leaf mines
Triphosa dubitata	G	IX–V	leaves
Xestia baja	G	IX–V	leaves
Yponomeuta evonymella	*	V–VI	leaves
Hymenoptera			
Pamphilius sylvaticus	R	VI–VII	leaves
Pristiphora retusa	*	VI–VII	leaves
Coleoptera			
Anthonomus bituberculatus	R	III–IX	shoots and buds
A. humeralis	R	IV–IX	shoots
Furcipus rectirostis	R	VII–IX	fruit
Magdalis ruficornis	R	IV–X	leaves
Phytodecta pallida	G	IX–VI	leaves
Rhynchites aequatus	R	IV–VIII	fruit
R. auratus	R	IV–VIII	fruit and leaves
R. pauxillus	R	IV–VII	leaves
Saperda scalaris	G	I–XII	dead wood and leaves
Scolytus rugulosus	R	I–XII	branches and stems

several species would be present at the same time, but only two overlap, *O. brumata* and *Y. evonymellus*. One species, *A. umbrana*, begins feeding when these finish, and another, *L. temerata*, starts feeding later still. The other three species are late summer feeders and finish feeding in May the following year. There is thus some evidence for temporal segregation

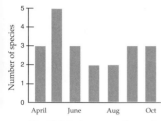

Fig. 14. Maximum numbers of different species of leaf-feeding Lepidoptera (excluding miners) found feeding on *Prunus padus* at any one time.

(fig. 14) which must reduce feeding pressure on the trees to a certain extent. In May, when the number of species feeding is highest, two of the species are probably reducing their feeding rate before pupation. Thus competition for resources has been reduced markedly by temporal separation.

Resource partitioning by spatial separation on a host plant is exemplified by the aphid species feeding on oaks in the British Isles (Dixon, 1985). The five species feed at different heights within the tree (fig. 15) and consequently avoid competition. Also, they differ greatly in size. As we saw when considering fractals, the smallest species, although feeding on the smallest-sized resource, has the greatest quantity of resources (trees having more leaves than twigs or branches) available to it and is one of the most abundant aphid species found on oak.

Distinct signs of resource partitioning are also seen among leaf-mining Lepidoptera on cherry trees, although these are competing for a single leaf as opposed to a whole tree. Of the four species of leaf miner that feed on *P. avium*, two feed on the leaf margins and the others feed on the upper and lower sides of the leaves respectively (table 3). In addition, the case-bearing leaf-miner *Coleophora hemerobiella* does not live within the leaf as the other three species do, but moves across the surface pausing periodically to feed on the internal contents of the leaf. On *P. cerasus*, there are only three species of leaf-miner and these appear to have partitioned the leaf between them in an ideal manner, one species being confined to the middle, another to the margins and the third species having the rest.

Further investigation may reveal resource partitioning of other types. For example, species may differ in their relative abundance at different heights in the tree, or on leaves of different age along a shoot, or in the sunnier or shadier parts of a tree. Sampling methods that might be used in this sort of investigation are described on page 66.

Fig. 15. The feeding positions of six species of aphids living on oak (a) and the relationship between body length and the length of the mouthparts plotted on a logarithmic scale (b). (After Dixon, 1985, with permission from Blackie.)

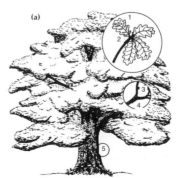

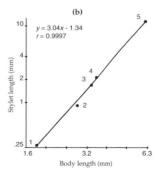

$y = 3.04x - 1.34$
$r = 0.9997$

Table 3. *Leaf-mining Lepidoptera feeding on* Prunus avium *and* Prunus cerasus.

Species	Feeding site	*Prunus avium*	*Prunus cerasus*
Coleophora hemerobiella	case-bearing; feeds on upper side of leaf	+	-
Ectoedemia atricollis	leaf margins	+	-
Phyllonorycter cerasicolella	underside of leaves	+	+
Stigmella prunetorum	leaf margins	+	+
Tischeria guanacella	middle of leaf	-	+

Table 4. *Percentages belonging to the main insect guilds among the insect species feeding on different species of cherry in Britain.* (Figures in brackets are the actual number of species in each guild.)

Guild	*avium*	*cerasifera*	*cerasus*	*laurocerasus*	*padus*
Chewers	60.0 (24)	75.0 (3)	71.8 (23)	100 (4)	80.7 (20)
Miners	10.0 (4)	0 (0)	9.4 (3)	0 (0)	3.8 (1)
Sap suckers	30.0 (12)	25.0 (1)	18.8 (6)	0 (0)	15.5 (4)

It was this form of spatial resource partitioning, as shown by the beetles feeding on the bark of *Pinus strobus* and the Lepidoptera chewing the leaves of *Prunus padus* and mining the leaves of *P. avium* and *P. cerasus*, that led biologists to the concept of guilds.

Guilds

A guild as defined by Root (1967) is a group of species that exploit the same class of environmental resources in a similar manner. Thus some of the Lepidoptera that we have discussed above belong to the leaf-mining guild. Other leaf-feeding guilds include pit feeders such as flea beetles which eat small patches of leaf, usually without perforating both surfaces, strip feeders or leaf chewers such as caterpillars which chew down the side of a leaf in strips, bud feeders such as winter moth caterpillars, which feed on or within the newly expanding buds, and sap feeders such as aphids and froghoppers, which feed on plant sap using piercing and sucking mouthparts. Several different guilds are represented among the insects feeding on *P. padus* (table 2). In this case, the leaf-chewing group is by far the largest guild with 21 species, but they are not all feeding on exactly the same resources. The sucking, mining and galling guilds are not particularly well represented. The proportions of species in the different guilds are not the same for all the cherry species, although the leaf-chewing group is always the largest guild (table 4). For a given species of cherry, the insect species representing the different guilds will differ from one patch of trees to another. When starting an investigation it is useful to compile a table of guild composition for the trees involved. The scope for inter-specific competition is expected to be greater within guilds than between them, so a guild with several species is of particular interest. Kirk (1992) gives a guild diagram for the insect species that feed on cabbages and oilseed rape.

Seasonality

There are marked seasonal changes in the chemical composition of leaves. For example, proteins, the building blocks of life, contain nitrogen, which is therefore one of the most important elements in nature. Protein is just as important to the plant as it is to the insect that feeds upon it. The amount of nitrogen within a plant is thus often used as a measure of the quality of that plant as food for insects. Plant nitrogen content can be expressed in a number of ways, for example as total nitrogen, soluble nitrogen, or amino acid or protein nitrogen (Strong and others, 1984). Which is most useful? The answer depends on what feeding strategy the insect herbivore adopts. Soluble nitrogen is what matters to a sap-feeding insect such as an aphid, whereas total nitrogen may be a more appropriate measure to use when investigating a chewing insect such as a caterpillar.

In the bird cherry *P. padus* soluble nitrogen levels peak in the spring, gradually decline until the beginning of autumn and then begin to rise again (fig. 16). Total nitrogen is also high in the spring but then declines throughout the remainder of the year with the minimum being reached in late summer and early autumn (fig. 16). Unlike soluble nitrogen, there is no late season peak. This is also true for the other deciduous cherries such as *P. avium*.

How do the insects feeding on these trees respond to these changes in the quality of the food provided by their host plant? Aphids illustrate those responses very well. Two aphid species that are closely associated with their cherry hosts are *Rhopalosiphum padi* on bird cherry and the cherry blackfly *Myzus cerasi* on gean. Both of these aphids spend the summer on a host plant species different from the one on which they overwinter. The bird cherry aphid migrates to grasses and cereals in the summer, whereas the cherry blackfly migrates to bedstraw, cleavers or speedwell. The winged, migratory aphids are produced in response to deterioration in the quality of their host plant in combination with crowding. Some aphid species can probably be induced to remain on their primary hosts for longer than normal either by increasing the nitrogen levels of the plant by fertiliser application or by reducing the number of aphids per colony. It is known that the black bean aphid *Aphis fabae* can remain on its primary host for the whole year if young leaves are available, so it would be worth investigating whether the aphids feeding on cherry trees also possess such a flexible life style.

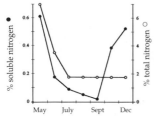

Fig. 16. Seasonal changes in leaf nitrogen content of *Prunus padus*. Black dots, left axis: percentage soluble nitrogen (after Dixon, 1973); circles, right axis: percentage total nitrogen (Leather, unpublished.)

ovariole: an insect has two ovaries, each of which is composed of a number of ovarioles within which are the eggs or nymphs waiting to be laid.

morph: aphids, and some other insects, occur in more than one form or morph. For example, aphids have winged and non-winged adults, the alate and apterous morphs respectively.

The reproductive output of an aphid is largely determined by the number of ovarioles within the reproductive tract. Whereas in Lepidoptera the number of ovarioles is constant for a species, in aphids the number varies not only within species, but also in some cases with morph. It is not determined by the size of the aphid. An aphid with numerous ovarioles can reproduce at a faster initial rate than one with fewer ovarioles (Dixon, 1985). Aphids with high ovariole numbers have fewer resources left over for other structures such as flight muscles, and are thus not so good at dispersing as aphids with fewer ovarioles (Leather, Wellings & Dixon, 1983).

We have seen that cherry trees are good food sources for aphids early in the season and it is to the advantage of the aphids feeding on them at that time to exploit the high quality of their primary hosts. Thus one would expect aphids with high numbers of ovarioles to be present at bud burst and early flushing. As can be seen this is indeed the case (fig. 17). As conditions deteriorate the amount of investment in reproductive structures decreases and more resources are available for flight and dispersal. The winged forms, which will emigrate to the secondary hosts, have fewer and fewer ovarioles as conditions worsen. In the autumn, as conditions worsen on the secondary host plants, winged forms (gynoparae) are again produced and these migrate back to their primary host to produce rapidly maturing offspring that lay eggs before leaf fall. There is thus a close link between aphid life cycles and the plants' seasonality.

Fig. 17. Changes through the season (and with morph sequence) in the numbers of ovarioles in the aphids *Rhopalosiphum padi* (squares) and *Myzus cerasi* (dots). (After Leather, 1987.)

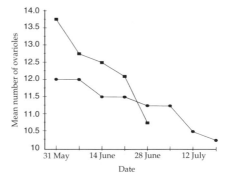

Induced defences

Some plants are more resistant or less susceptible to insect attack than others of the same species. This variability may be associated with differences in plant chemistry or in anatomical features such as hairy leaves or a tough cuticle. Resistance that is a permanent feature of an individual plant is known as constitutive resistance.

Since the 1970s evidence has accumulated suggesting that insect attack can somehow make the attacked plants more resistant to further attack by the same and other species of insect. This may be a short-term response (rapidly-induced resistance), in which chewing on a leaf produces a rapid chemical change in the plant, causing the insect to stop feeding at that site within minutes or hours. This rapidly-induced resistance is commonly cited as the reason why leaves often have lots of small holes in them rather than being eaten all in one go (Edwards, Wratten & Parker, 1992). Alternatively, the plant's response to insect feeding can be slower and longer-lasting. This delayed induced resistance can adversely affect the growth rates and survival of the larvae feeding on the damaged foliage in the same year or even in the year following the damage (Haukioja, 1991). This has been demonstrated in a number of different examples and continues to be a source of great debate. It has even been suggested that some plants when attacked by insects produce highly volatile chemicals that are carried in the air to other plants of the same species, which then arm themselves chemically in readiness for attack (Baldwin & Schultz, 1983; Rhoades, 1983). The evidence for this tree-to-tree communication has been challenged on more than one occasion in the so-called 'talking tree' debate (Fowler & Lawton, 1985), but recent studies on the biochemistry of induced responses makes such chemical communication more plausible (Karban & Baldwin, 1997).

So far there is little evidence for any of these induced responses in cherry trees although some preliminary work suggests that defoliation of bird cherry can result in chemical changes that make the next crop of leaves less suitable for caterpillars of the autumnal moth *Epirrita autumnata* in Finland (Neuvonen, Haukioja & Molarius, 1987) and reduce general insect attack in Britain (Leather, 1992).

It is surprising that so little is known about the induction of chemical defences in cherry trees because they contain a wealth of alkaloids which could provide a useful armoury against herbivores. This is an area in which further studies could be very rewarding. For example artificial damage to leaves made with a hole punch may affect their subsequent colonisation or damage by insects. It is also possible that further insect feeding may depend on the degree of damage, for example the number of holes per leaf. There is some evidence to suggest rapid communication between different parts of a plant. Damaging one leaf may cause the next leaf to become resistant to insect attack, even if it is undamaged. However, in some plant species damaging one leaf may have no effect on its neighbour. Defoliating whole trees or saplings also affects their subsequent colonisation by different insect groups and the degree of insect damage that they sustain. Probably some insect groups are more affected by defoliation treatments than others. Research on the induction of chemical defences is likely to be very productive and controversial.

Indirect or plant-mediated inter-specific competition

The plant's response to defoliation may be used by an insect species to compete with another species feeding on the same tree. Two species may influence each other even if they are not present on the same resource at the same time. There is some evidence for such indirect competition on bird cherry, between the bird cherry aphid *Rhopalosiphum padi* and the bird cherry ermine moth *Yponomeuta evonymella*. Although these two species feed at the same time of year and frequently occur on the same trees, the competitive effects are indirect and are mediated by the plant (Leather, 1988).

Individual trees of *P. padus* differ from one another in their suitability as food sources for *R. padi* and these differences are consistent from year to year. Those trees with high aphid numbers one year will have high aphid numbers the following year (Leather, 1986). A study of *P. padus* trees in Finland (Leather & Lehti, 1982) showed that trees with large numbers of overwintering aphid eggs had low numbers of *Y. evonymella* overwintering larval shields and vice versa. The same relationship is also seen in Scotland. The overwintering eggs of *R. padi* are in the angle between the bud and the shoot; the larval shields of *Y. evonymella* on the other hand are deposited on the shoot itself. Since the larval shield is on average 13 mm long and the moth is also 13 mm long the moth requires a space at least 13 mm long and probably up to 20 mm long in which to lay her eggs comfortably. Thus, the moth will favour trees on which the buds are widely spaced, whereas the aphid will favour those with a high density of buds on the shoot. Trees attacked by the moth always re-leaf, and in the autumn these replacement leaves senesce later than those on undefoliated trees and are thus less attractive to the aphids, which require a food source rich in soluble nitrogen. Senescent leaves contain relatively high levels of soluble nitrogen as the plant is mobilising its reserves prior to storage over the winter. Trees that are heavily attacked by aphids have more buds per cm of shoot and a reduced internodal distance (table 5). Thus either the feeding action of the aphid makes the tree less suitable for the moth – a clear example of indirect inter-specific competition – or the moths and aphids colonise different trees in the first place without necessarily having any effect on each other. Experimental work using clonal material has recently shown that it is indeed the type of defoliation that causes the differential colonisation, and that indirect competition is taking place.

larval shield: females of *Yponomeuta evonymella* lay their eggs in batches of about 50–100. The females cover the eggs with a secretion that hardens to form a resistant shield over the egg mass. The eggs hatch shortly after being laid, but the larvae remain beneath the shield during the winter, hence the term larval shield.

Experiments of this type can be set up fairly easily and should provide interesting insights into why certain insect species are restricted to certain trees. It would be very useful to extend this study to other pairs of insect species or guilds, particularly species that are relatively common. Correlations, positive or negative, will give a clue as to whether some form of competition or otherwise is occurring; but a manipulated experiment such as that described above will be needed to identify the exact competitive mechanism. Some examples of relevant techniques are given in Chapter 5.

Table 5. *Effects of different levels of aphid infestation (mean number of aphids per leaf ± standard error) on shoot characteristics of bird cherry trees.* (Data from Leather, 1988.)

	Infestation level		
	low	medium	high
	< 0.1 aphids per leaf	2–3 aphids per leaf	10 aphids per leaf
Buds per cm of shoot	0.59 ± 0.06	0.61 ± 0.07	0.73 ± 0.08
Mean internodal distance (cm)	2.17 ± 0.16	2.09 ± 0.16	1.77 ± 0.14

4 Identification

Identifying cherries

Cherries are very noticeable, especially in the spring when they are in bloom, and they are relatively easy to recognise as cherries. The native species are fairly distinct from each other. Cultivated forms are more difficult to name, but should not present problems if obtained from a garden centre. Species can be subdivided into subspecies or, if the different types are the result of artificial selection (for example for flower colour or shape), into cultivars (cultivated varieties).

Cherries belong to the genus *Prunus*. If your specimen is a species of *Prunus* it will have the following characteristics. The leaves are simple (undivided), and arranged alternately along the stem or in clusters on short side shoots. The flowers are radially symmetrical with five separate white or pink petals (or more than five if the flowers are double), and at least ten separate anthers (except in some double forms). The flowers are perigynous, and have a single style and carpel. The fruit is succulent and has a single stone. Peaches and almonds can be distinguished from other species of *Prunus* by their hairy ovary and fruit. They are not included in the key below. This is based on the key in Stace (1997), where the species of *Prunus* are described in more detail.

a perigynous flower:
a flower in which the petals, sepals and anthers are inserted on the rim of the extended receptacle, a bowl-shaped structure originating below the ovary (I.4, I.5)

I Cherries and related species of *Prunus*

1   Leaves not tough and leathery, deciduous (falling off in autumn)   2
–   Leaves tough and leathery, evergreen   12

2   Flowers solitary or in clusters of up to ten, not in elongate spike-like racemes   3
–   Flowers in groups of more than ten in elongate spike-like racemes   11

3   Flowers solitary or in groups, originating very close together or along a very short axis; no green bracts on the axis (but there may be some green bud scales at the base of the flower cluster)   4
–   Flowers arranged along an axis bearing green bracts near its base   *P. mahaleb*

4   Base of flower cluster with large (over 5 mm) bud scales which are often green or reddish   cherries 5
–   Base of flower cluster with no bud scales or with small (up to 3 mm) brown ones   plums 9

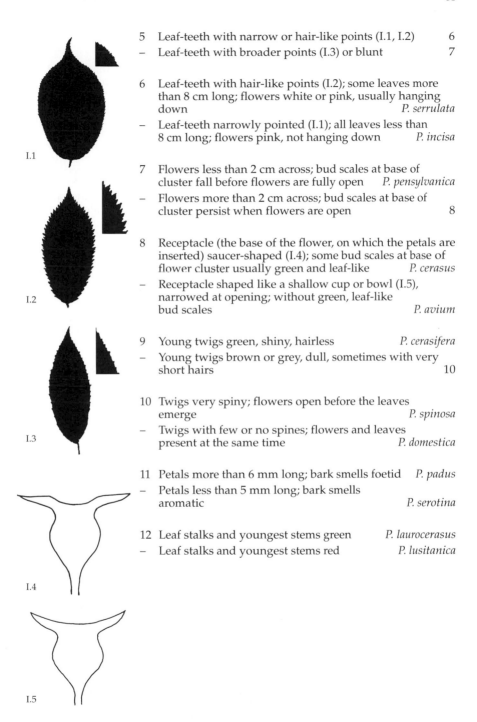

5   Leaf-teeth with narrow or hair-like points (I.1, I.2)   6
–   Leaf-teeth with broader points (I.3) or blunt   7

6   Leaf-teeth with hair-like points (I.2); some leaves more
    than 8 cm long; flowers white or pink, usually hanging
    down   *P. serrulata*
–   Leaf-teeth narrowly pointed (I.1); all leaves less than
    8 cm long; flowers pink, not hanging down   *P. incisa*

7   Flowers less than 2 cm across; bud scales at base of
    cluster fall before flowers are fully open   *P. pensylvanica*
–   Flowers more than 2 cm across; bud scales at base of
    cluster persist when flowers are open   8

8   Receptacle (the base of the flower, on which the petals are
    inserted) saucer-shaped (I.4); some bud scales at base of
    flower cluster usually green and leaf-like   *P. cerasus*
–   Receptacle shaped like a shallow cup or bowl (I.5),
    narrowed at opening; without green, leaf-like
    bud scales   *P. avium*

9   Young twigs green, shiny, hairless   *P. cerasifera*
–   Young twigs brown or grey, dull, sometimes with very
    short hairs   10

10 Twigs very spiny; flowers open before the leaves
    emerge   *P. spinosa*
–   Twigs with few or no spines; flowers and leaves
    present at the same time   *P. domestica*

11 Petals more than 6 mm long; bark smells foetid   *P. padus*
–   Petals less than 5 mm long; bark smells
    aromatic   *P. serotina*

12 Leaf stalks and youngest stems green   *P. laurocerasus*
–   Leaf stalks and youngest stems red   *P. lusitanica*

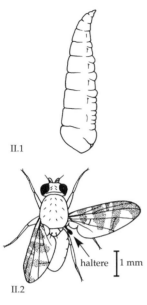

II.1

II.2

haltere |1 mm

II Major groups of insects

These keys deal only with insect species that depend on
cherry trees because the larvae or adult stages feed on cherry.
Very many other insect species visit cherry trees and are
found there, perhaps resting or seeking prey on the leaves, or
attending aphid colonies, or taking nectar and pollen from
the flowers. The keys should work reliably for insects seen
feeding or laying eggs on cherry tissues, but other insects,
especially winged adults which may be casual visitors to
cherry, may key out wrongly as if they were cherry-
associated species. It is therefore necessary to check very
carefully against the description and illustrations and, to be
absolutely confident, against the more detailed taxonomic
works listed at the head of each key.

Although many cherry insects can be identified using the
unaided eye or a x 10 lens, a dissecting microscope is needed
for the Hemiptera: Homoptera (Key III), and is helpful for
some other groups. It may sometimes be possible to borrow
one from a local school or college.

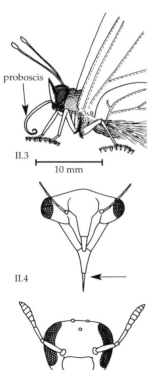

proboscis

II.3

|———| 10 mm

II.4

II.5

1   Without legs; maggot-like with no distinct head (II.1)
    fly larvae DIPTERA
Key VIII, p. 58

–   With three pairs of jointed legs   2

2   With wings   3
–   With no wings (or only wing buds)   8

3   Only one pair of wings; rear pair modified into
    club-shaped halteres (II.2)   adult flies DIPTERA

    The majority of flies found on cherry will be using the tree for nectar or
    shelter, so a key to adult flies is not included here (see Unwin, 1981;
    Colyer & Hammond, 1968 for identification). Flies reared from cherry
    can be identified from the larval stages.

–   With two pairs of wings; front pair may be hard
    and thick   4

4   Both pairs of wings of similar texture   5
–   Front pair of wings partly or wholly hardened and
    thickened   7

5   Wings covered with dust-like scales that rub off onto the
    fingers; mouthparts forming a coiled proboscis (II.3)
            moths and butterflies LEPIDOPTERA Key V, p. 48
–   Wings without scales; mouthparts not like this   6

6   Head with piercing mouthparts (II.4)
            bugs HEMIPTERA: Homoptera Key III, p. 34
–   Head with well-developed jaws (II.5)
            sawflies HYMENOPTERA Key VII, p. 57

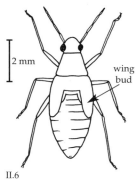

2 mm

wing bud

II.6

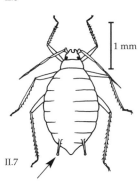

1 mm

II.7

7 Forewings hard and thick all over
                        beetles COLEOPTERA Key IX, p. 60
– Forewings mostly hard and thick but membranous
   towards the tip   bugs HEMIPTERA: Heteroptera
                                                    Key X, p. 64

8 Body covered with scales
                        wingless female moths LEPIDOPTERA Key V, p. 48
– Body without scales   9

9 With piercing mouthparts (II.4)   10
– With biting jaws (II.5)   11

10 Wing buds present (II.6); no obvious dorsal
    projections at rear   immature bugs HEMIPTERA
                                                    (not keyed out here)
– No trace of wing buds; a pair of dorsal projections
   (siphunculi) at rear (II.7)
                        aphids HEMIPTERA: Homoptera Key III, p. 34

11 Ant-like, with a distinct waist
                        ants HYMENOPTERA, probably visiting aphids
                                                    (see Skinner & Allen, 1996)
– Caterpillar-like, without a waist   12

12 Rear end of caterpillar with a pair of clasping,
    sucker-like feet, each of which ends in a circle of minute
    hooks (or crochets)
                        caterpillars of butterflies and moths LEPIDOPTERA
                                                    Key IV, p. 39
– Rear end without clasping feet; any tube-like feet in this
   position lack crochets
                        larvae of sawflies HYMENOPTERA Key VI, p. 56
                        or beetles COLEOPTERA (not keyed out here)

III Adult homopteran bugs Hemiptera: Homoptera

For critical identification, particularly of species not listed as cherry specialists, see Blackman & Eastop (1994), Stroyan (1984) and Heie (1986, 1992, 1994).

Aphid life cycles are complex. In the autumn, winged male and female aphids mate and seek out their primary host plant on which the female lays her eggs. In spring these develop into further winged forms. These fly off and settle on another plant species, the secondary host, where the species spends the summer. On the secondary host, all nymphs develop into wingless female adults that reproduce without mating (that is, by parthenogenesis). Aphids of this summer generation do not lay eggs but give birth to living young (they are viviparous). In the autumn the wingless parthenogenetic females start producing a winged sexual generation of aphids that migrate back to the primary host. Blackman (1974) gives more details.

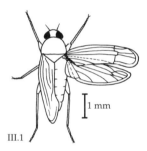

III.1

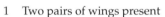

pterostigma

III.2

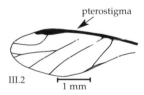

1 mm

III.3 CCS IAV

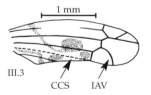

III.4 CCS IAV

1   Two pairs of wings present   2
–   Wings entirely absent (family Aphididae, immature or wingless viviparous female aphids)   10

2   Forewing without a thick, dark area (pterostigma) on front edge, and with at least two veins parallel to front edge (III.1)   family Cicadellidae (8 species) 3
–   Forewing with a pterostigma (III.2) and only one vein, or no veins, parallel to front edge
      family Aphididae (winged aphids)

A few of these are easy to separate (for example *Hyalopterus pruni* and *Brachycaudus helichrysi*) but most are difficult to identify without making microscopic preparations. As a colony with winged viviparous females always contains some immatures or wingless viviparous females as well, winged viviparous females are not keyed here. Enthusiasts can take them further using Stroyan (1984) and Heie (1986, 1992, 1994).

3   Forewings with distinct border around the end (III.1)
      Alebra wahlbergi

Length 3.0–4.5 mm. In the male, which is smaller than the female, head yellowish, pronotum and forewings deep orange. Female with yellowish head and pronotum, latter sometimes streaked with orange; forewings broadly yellow, pink or orange along rear edge and usually streaked with yellowish or orange along centre. Both adults and larvae suck juices out of leaves of a wide variety of trees. *Adults* July-September. Widespread in England and Wales as far north as south Lancashire.

–   Forewings without such a border around the end   4

4   Internal Apical Vein (IAV) of forewing curving towards hind margin of wing so that it reaches the margin nearer to the corioclaval suture (CCS) than to the wingtip (III.3) 5
–   IAV of forewing straight, reaching margin near wingtip (III.4)   9

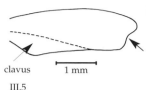

clavus   1 mm

III.5

5 Tip of forewing looking as if a bite has been taken
out of it (III.5)   *Aguriahana stellulata*

Length 4.0–4.5 mm. Head and pronotum whitish, slightly tinged
brownish or yellowish. Forewings whitish, with tip excised and
black-brown blotch at wing tip and at tip of rear margin. Only recorded
on cherry, plum and lime. *Adults* July to October. Widespread in
southern half of England and in Wales.

– Tip of forewing rounded normally   **6**

6 Forewing with well-defined brick-red or orange spots,
especially near CCS (III.3)   *Typhlocyba quercus*

Length 3.0–3.5 mm. Head and pronotum yellowish. Forewings with
basal two-thirds strongly marked with red or orange spots, while apical
part has veins outlined in brownish-grey. Both adults and larvae feed
by sucking juices from leaves of a wide variety of trees. *Adults* July to
October. Throughout the British Isles.

– Forewing pale yellowish and without such spots   **7**

7 Forewing with clavus (labelled on III.5) dark along inner
margin   *Edwardsiana crataegi*

Length 3.5–4.0 mm. Head and pronotum pale yellowish; forewings pale
yellowish. Both larvae and adults suck juices from leaves of a variety of
rosaceous trees. *Adults* July to September. Throughout the British Isles.

– Forewing without dark markings on clavus   **8**

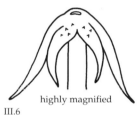

highly magnified

III.6

8 Aedeagus of male with four branches at the tip (use a
microscope) (III.6)   *Edwardsiana rosae*

Length 3.5–4.0 mm. Superficially similar to the following species;
females cannot be distinguished. Both larvae and adults suck juices
from leaves of a range of rosaceous trees. *Adults* June and again August
to October. Throughout England and Wales, just entering south
Scotland.

– Aedeagus of male with six branches at the tip (III.7)
                                                    Edwardsiana prunicola

Length 3.5–4.0 mm. Superficially similar to the previous species;
females cannot be distinguished. Larvae and adults feed on a variety of
rosaceous and other trees. *Adults* July to October. Throughout the
British Isles.

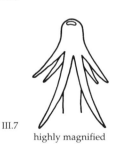

III.7

highly magnified

9 Forewing uniform yellowish without reddish
or pinkish markings   *Alnetoidia alneti*

Length very variable. Head, pronotum and forewings uniformly
yellowish and unmarked. Both larvae and adults feed by sucking juices
out of the leaves of a wide range of trees. *Adults* June to October.
Throughout the British Isles.

– Forewings with red or pinkish orange band
along corioclaval suture   *Zygina flammigera*

Length 3.0–3.5 mm. Head variable; pronotum normally with two
longitudinal red bands, forewings with zig-zag red band along rear
edge. Feeds on various trees and bushes. *Adults* July to May,
overwintering in conifers, holly and ivy. Throughout England, Wales
and Scotland.

10 Enclosed within red and yellow leaf galls
Myzus padellus

The *wingless aphids* are yellow with dark siphunculi, head and caudal process. The *winged females* are pale with a small darker patch on the back and some mottling on the sides. The head, siphunculi and caudal process are dark. The primary host is *Prunus padus* on which colonies develop in red and yellow leaf galls in spring. The secondary host is hemp-nettle (*Galeopsis*).

–   Only partially enclosed by leaf tissue or not at all   11

11 Black and shiny all over   *Myzus cerasi*

The *wingless aphids* are shiny black or dark brown with dark siphunculi, antennae, femora and tarsi. The *winged females* have a darkish abdomen with a large dark patch on top, often extending into the marginal plates and with cross bands or spots in front of it and behind it. The siphunculi are dark. The primary host is *Prunus cerasus* or *P. avium* but other cherries are also attacked. *P. avium* responds to the presence of the aphids in the spring with much curling and contortion of the leaves to form a nest of leaves. On *P. acida* and *P. serralata* the leaves are deformed only slightly or not at all by the presence of the aphid. This difference in response of the primary host plant has led to the suggestion that two species of aphid are involved, but this is not widely accepted as the aphids on the different hosts are morphologically identical. The colonies are visited by ants. The secondary hosts are usually bedstraws, eyebrights or speedwells.

–   Not entirely black and shiny   12

12 Green or reddish with shiny black upper surface
Brachycaudus cardui

The *wingless aphids* are pale green or reddish with black siphunculi and with a shiny blackish upper surface in the adults, but not in the nymphs, which consists of separate bars across the thoracic segments, and a black shield on the abdomen. The primary host is usually peach, plum or blackthorn, but this aphid occasionally ventures on to cherry, where the first generation causes curling of infested leaves in the spring. The aphids are visited by ants. They move to the secondary hosts, mostly herbaceous Compositae or Boraginaceae, in June.

–   With less extensive dark markings, or none   13

13 Brown   immatures of *Brachycaudus helichrysi*

Immatures or first-generation females are brown. *Wingless females* of the second generation are green, yellow or yellowish white with the upper surface unpigmented and the siphunculi pale. The tips of the antennae and legs are dusky. The primary host is usually plum or blackthorn but cherry is occasionally used. In spring the infested leaves of the host tree become curled and are visited by ants. The secondary host may be any of a wide range of herbaceous plants, especially Compositae.

–   Pale, greenish or yellowish   14

14 Covered with a whitish waxy powder   *Hyalopterus pruni*

The *wingless aphids* are green, covered with a waxy powder which makes them appear pale green. The tips of the antennae, tibiae, tarsi and siphunculi are brownish (pl. 2.3, 2.4). The *winged females* are similar, but the head and thorax are blackish and the abdomen sometimes has faintly dusky markings. The primary host is usually plum, blackthorn, peach or almond, although this aphid occasionally ventures onto cherry. On the primary host, little or no distortion of the leaf occurs unless the infestation is very heavy, but the leaves are much fouled with honeydew as the aphids are not visited by ants. In mid June the aphids migrate to their secondary host, the common reed *(Phragmites australis)* or occasionally other species of grass, where they feed until late September on the upper surface of the leaves.

– Without waxy powder   15

15 Pale green with reddish spots   *Rhopalosiphum padi*

The *wingless aphids* are light green with reddish spots at the bases of the siphunculi. The siphunculi themselves are pale with dark tips (pl. 2.1). The *winged females* are olive green with brown blotches. The head, thorax and siphunculi are black. The primary host is *Prunus padus* and occasionally other species of *Prunus*. Eggs hatch in early April. The first generation in spring live on the undersides of the leaves causing them to roll or fold downwards longitudinally (pl. 2.1). The aphid colony is thus enclosed in something akin to an open gall. The colonies are visited by ants. Winged females produced in these colonies fly to the secondary host in June, although occasional colonies can sometimes still be found on *P. padus* in July. The secondary hosts are usually various grasses, including cereals, on which the following generations develop in the sheaths of the lower leaves. As the aphids can be vectors of virus diseases, especially Barley Yellow Dwarf Virus, they can be very harmful to cereal crops. Migration back to *P. padus* for egg-laying usually takes place in September or October, although the mild winters of southern England allow some females to overwinter on the secondary host (Stroyan, 1984).

– Without reddish spots   16

16 Greenish or yellowish with a green line along the
   middle   17
– Aphid greenish or yellowish but without a line
   along the middle   18

17 With green bands across the body as well as the
   green line along the middle   *Phorodon humuli*

The *wingless aphids* are whitish, greenish or light green with a green stripe along the back and short green bands across the body. The *winged females* are similar, but winged. The primary host is usually plum or blackthorn but cherry is also used. The secondary host is hop (*Humulus lupulus*). Interestingly, this was the aphid species in which alternation between primary and secondary hosts was first discovered. On the primary host only slight curling of the young leaves occurs and little or no harm is done; however, this aphid can be a serious pest of hop. It is not visited by ants. Migration from *Prunus* to hop occurs in June–July. The aphids occasionally stay on *Prunus* all summer, but when this happens they fail to produce males. Eggs are laid on *Prunus* in October.

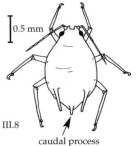

0.5 mm

III.8

caudal process

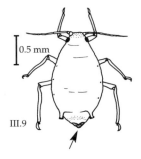

0.5 mm

III.9

– Without green bands across the body
Rhopalosiphum insertum

The *wingless females* are light green with a green stripe along the middle. The siphunculi are pale with dark tips. The *winged females* have a blackish head, thorax and siphunculi. The abdomen is green with brown blotches. The primary host is usually apple or hawthorn, but cherry is very occasionally used. Eggs hatch in April and the aphids cause curling of the leaves of the primary host. The colonies are visited by ants. Winged females fly to the secondary hosts, usually various grasses, in late May or early June and form colonies on the roots and basal parts of the stem. Occasionally colonies are still found on the primary host in July. Migration back to the primary host by winged oviparous females occurs in September or October.

18 With a distinct caudal process (III.8)   *Myzus persicae*

The *wingless aphids* are yellowish green, green or pale yellowish. The first generation female or fundatrix is often partly reddish. The tips of the antennae and tibiae are somewhat darkened as also are the siphunculi occasionally. The *winged females* have a dark brown head and thorax. The abdomen is yellowish green to yellow with dark brown or black patches and spotting on the upper surface. The tibiae and siphunculi are dark, especially towards the tip. In the nymphal stage the body is reddish. The species is sometimes called the green peach aphid as the primary host is peach, *Prunus persica*. This aphid has also been found on other species of *Prunus*, including *P. serotina* and *P. serrulata*. The secondary hosts are a very wide range of herbaceous plants from many different families. These hosts include many agricultural crops, such as potato, beet, kale or cabbage and flowers such as tulips and carnations. As this aphid can be a vector for more than a hundred plant viruses, including Potato Leaf-roll and Beet Yellows, it is regarded as the most harmful aphid species in the world.

– Aphid with an insignificant caudal process (III.9)
Brachycaudus helichrysi
(see couplet 13 above)

IV Caterpillars of Lepidoptera

For further identification see Buckler (1886–1910), Haggett (1981), Carter & Hargreaves (1986), Stokoe & Stovan (1948) (for the larger moths, the Macrolepidoptera); Bradley, Tremewan & Smith (1973, 1979) (for Tortricidae); Heath & Emmet (1976–1996).

1   Larva feeding within the leaf tissue in a 'mine'   2
–   Larva feeding in some other way   16

2   At least initial part of mine in the form of a linear gallery   3
–   Mine wholly in the form of a blotch   5

3   Empty egg-shell, at origin of mine, within the leaf and so concealed (use a lens)
      family Lyonetiidae (1 species) *Lyonetia clerkella*

Larva forms a long linear gallery often more than 10 cm long, with the black excreta in a narrow line down the middle. The mine often crosses the leaf veins and so sometimes causes areas of leaf to die and fall out or to change colour, especially in the autumn. There may be three generations in a year in southern Britain, and two in the north.

–   Empty egg-shell, at origin of mine, on the leaf surface and in the form of a flattened disc (use a lens)
      family Nepticulidae (2 species)   4

4   Mine forming a gallery throughout, although much contorted initially   *Stigmella prunetorum*

Initial gallery much convoluted and forming a brownish false blotch; later gallery straighter. Mines are almost filled with arcs of excreta. Two generations: July and September to October.

–   Mine starts as a linear gallery but then expands into a blotch   *Ectoedemia atricollis*

Initial gallery is filled with scattered brown excreta and often follows the margin of the leaf. The mine then expands into a large blotch in which the excreta are randomly scattered.

5   Larva lives in a case while mining the leaf, retreating into the case when disturbed
      family Coleophoridae (5 species) 6
–   Larva residing entirely within the mine   13

6   Case small (early), less than 4 mm high   7
–   Case large (later), more than 4 mm high; either cigar-shaped or pistol-shaped   11

7   Case straight, made from a single piece of leaf   8
–   Case angled or curved   9

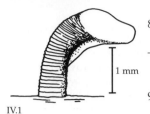

IV.1

8   Case extended with silk at hind end and with a
     ventral ridge or keel   *Coleophora trigeminella*
–   Case not extended and without a ridge or keel
                                              Coleophora adjectella

9   Upright portion of case smooth and black
                                              Coleophora anatipennella
                                              (see couplet 11)
–   Upright portion of case, near point of attachment to
     substrate, rough and made of distinct rings of
     leaf cuticle and silk   10

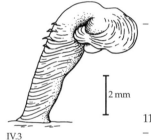

IV.2

10   Case pennant-shaped, with the two-valved pennant part
     formed from a pair of oval pieces of leaf-cuticle attached
     to a shaft at 45°. The shaft, about 2 mm long when fully
     formed, is composed of stacked rings of leaf-cuticle held
     together with silk (IV.1)   *Coleophora coracipennella*
                                              C. prunifoliae
                                              C. spinella
                                              (see couplet 12)
–   Case shaped like a walking-stick handle. Constructed in a
     similar way to the previous species, but the shaft is
     longer and curved more at the hind end so that the two-
     valved portion points directly towards the point of
     attachment) (IV.2)   *Coleophora hemerobiella*
                                              (see couplet 12)

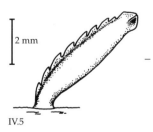

IV.3

11   Case brown and cigar-shaped   12
–   Case black and pistol-shaped (IV.3)
                                              Coleophora anatipennella
     The fully formed case is pistol-shaped, shiny black and about 7 mm
     high. The barrel of the pistol is attached to the substrate at 70–80°.

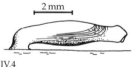

IV.4

12   Three options:
–   Case 5–6 mm long, dark red-brown and lying
     horizontally on substrate (IV.4)   *Coleophora trigeminella*
–   Case 6–7 mm long, dark red-brown and projecting at 45°
     to substrate (IV.5)   *Coleophora adjectella*
                                              C. coracipennella, C. prunifoliae, C. spinella

     All four species are reported from *Prunus cerasus* but *C. coracipennella* is
     also reported from *P. padus* and ornamental *Prunus* species.
     Unfortunately the four species cannot currently be separated in their
     larval stages and the adults can only be separated on genitalia
     characters.

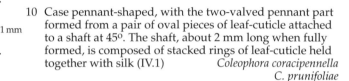

IV.5

–   Case 8–10 mm long, dark brown and projecting at
     90° to the substrate (IV.6)   *Coleophora hemerobiella*

     In Britain reported only from *Prunus avium* so far. The initial small two-
     valved case is formed in September. The larva hibernates in the case
     until the spring when it adds to the case while feeding actively from
     April to June. At this point the full walking-stick-shaped case is
     completed. The larva is then inactive, without feeding, for the rest of
     the summer (aestivating) and over winter (hibernating) until April of
     the second year, when the cigar-shaped case is formed. Feeding stops in
     June and after pupation in the case the adult emerges in July.

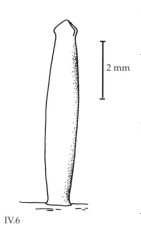

IV.6

13 Mine distorting the leaf; as the larva grows, it spins silk
    onto the underside of the leaf cuticle which then
    contracts and causes a blister-like mine   14
 – Mine remains flat and does not distort the leaf
                       the beetle *Trachys minutus* (Coleoptera)
                    (see key IX, couplet 3 or early stages of the species
                                        keyed out in couplet 14)

14 Mine on the upper surface of the leaf
              family Tischeriidae (1 species) *Tischeria gaunacella*
    The mine starts as a short greenish gallery but is quickly incorporated
    into a blotch. The leaf edge often curls over and conceals the mine. The
    larva mines in September and October but pupates within the mine and
    adult emerges the following spring. Not seen in Britain since last
    century.
 – Mine on the underside of the leaf
                        family Gracillariidae (2 species) 15

15 Blister mines on *Prunus padus*   *Phyllonorycter sorbi*
    Two generations, larvae feeding from June to July and September to
    October. The second generation pupates in the mine in October and
    adults emerge the following spring.
 – Blister mines on *Prunus avium, P. cerasus* or
    cultivated cherry   *Phyllonorycter cerasicolella*
    Two generations, larvae feeding from July and September to October.
    Second generation larvae remain in the mine throughout winter and
    pupate in April.

16 Larvae entirely concealed either within the plant or
    within a spinning   17
 – Larvae more or less exposed   28

17 Larvae concealed and feeding within a spun shoot or leaf
                             family Tortricidae (11 species) 18
 – Larvae concealed by burrowing into the plant tissues   25

prothoracic plate

anal plate

2 mm

IV.7

18 Larva with reddish brown body   *Epiblema trimaculana*
    Larva: head light to dark brown; prothoracic plate (IV.7) brown to black;
    body reddish brown sometimes tinged with green, paler below; anal
    plate (IV.7) light to dark brown. Feeds amongst spun leaves in April
    and May, moving from shoot to shoot. Pupates in loosely spun cocoon
    in larval feeding place. *Adult* emerges in May.
 – Larva with body differently coloured   19

19 Larva with yellowish white body   *Epinotia signatana*
    Larva: head yellowish brown to black; prothoracic plate pale brownish
    yellow; body yellowish white. Feeds in May in a folded leaf or spun
    shoot. Pupates in larval feeding place. *Adult* emerges in June.
 – Larva with green to yellow-green body   20

20 Larva with yellowish brown head   21
–   Larva with head otherwise coloured   22

21 Larva with prothoracic plate (top of the segment
    bearing the first pair of legs) black   *Acleris rhombana*
                                             rhomboid tortrix

Larva: prothoracic plate black, becoming browner as the larva ages.
Body light green to yellowish green; anal plate green. Feeds on the
buds initially but later in a spinning of leaves or flowers; April to June.
Pupates in the last larval spinning or between spun leaves. *Adult*
emerges in August.

–   Larva with prothoracic plate similar in colour
    to the head (yellow-brown)   *Adoxophyes orana*
                                           summer fruit tortrix

Larva: head and prothoracic plate initially black but becoming
yellowish brown in later instars. Abdomen variable, yellow-green,
olive-green or dark green; anal plate the same colour as abdomen. Two
generations of larvae, June to July and September to May. The first
generation initially skeletonizes leaves from below, working from
under a white protective web, but older larvae live in spun leaves. Such
spinnings often include fruit, which is thus damaged by the larva
feeding. The second generation in its early instars also skeletonizes
leaves. It overwinters in the second or third instar in a silken
hibernating chamber in any suitable niche on a tree, such as in the fork
of a twig. In late March the larva begins to feed again, eating into buds
and developing leaves. Pupation occurs in the larval feeding place and
the pupal stage lasts about 10 to 20 days. *Adults* emerge in June and
August.

22 Head of larva greenish with darker patches   23
–   Head black to black-brown   24

23 Head greenish with conspicuous black stripe
    at the sides behind   *Pandemis cerasana*
                                           barred fruit-tree tortrix

Larva: head pale green to brownish green with a conspicuous black
stripe at the side behind; prothoracic plate pale green to yellowish
green, marked with black at the sides. Body pale green, darker above;
anal plate pale green to yellowish green, spotted with black. Some
larvae hatch in September, overwinter in silken hibernacula on a twig
and begin feeding on the buds in spring. Other larvae do not hatch
until spring. All later feed in a folded or rolled leaf until May. They
pupate in the larval feeding place or in a folded leaf. *Adult* emerges in
June.

–   Head greenish, shaded with yellow-brown, but without a
    black stripe at the sides behind   *Pandemis corylana*
                                           chequered fruit-tree tortrix

Larva: head, prothoracic plate and thoracic legs predominantly green;
head with patches of yellowish brown and prothoracic plates lightly
marked with black. Body green; anal plate green. Feeds in May to July
by spinning leaves together or folding a leaf longways. Pupates in
larval feeding place. *Adult* emerges in July.

24 Three options:
- Body of larva bright green   *Hedya pruniana*
                                       plum tortrix

Larva: head, prothoracic plate and thoracic legs black. Body bright green; anal plate greenish brown or black. Feeds in April and May, at first eating out the heart of young shoots and then feeding in a folded leaf or between two leaves spun together with silk. Pupates in larval feeding place or in a spun leaf. *Adults* emerge in May and June.

- Body of larva grey-green to dark olive-green
                                       Hedya dimidioalba
                                       marbled orchard tortrix

Larva: head, prothoracic plate and thoracic legs black. Body grey-green to dark olive-green; anal plate blackish brown. Hatches in August and passes the winter as a very young larva in a silken hibernaculum in a crevice in the bark. Begins feeding again in spring in spun flowers or leaves until late April or early May. Pupates in larval feeding place or between two spun leaves. *Adult* emerges in June.

- Body of larva light green   *Acleris notana*
                                       or *Acleris umbrana*

Acleris notana
Larva: head, prothoracic plate and thoracic legs black. Body light green. There are two generations in the south, feeding in May to June and August to September but only a single generation in the north (June to July). Larvae feed within a spinning of two or three leaves. *Adults* emerge in July (south) or August (north) and again in October in the south.

Acleris umbrana
Larva: head, prothoracic plate and thoracic legs black. Body light green. Feeds from spun leaves in June and July. *Adult* emerges in August.

25 Larvae concealed by burrowing into bark
                          family Tortricidae *Enarmonia formosana*

Larva: head brown; prothoracic plate translucent in front, becoming brown behind. Body shining translucent grey; anal plate shining grey speckled with dark grey. Feeds September to May in tunnels in the bark. Its presence in the bark is revealed by piles of extruded red-brown frass usually in rough-barked areas on the trunks of older trees. *Adult* emerges in June.

- Larvae concealed by burrowing into shoots,
  leaf stalk or fruit   26

26 Larvae concealed by burrowing in shoot or leaf stalk   27
- Larvae concealed by burrowing into flesh of fruit
                          family Tortricidae *Pammene* or *Cydia* (4 species)

The species cannot be reliably separated on gross characters of larvae, and are best identified by rearing adults.

Cydia funebrana (plum fruit moth)
Larva: head dark brown to black; prothoracic plate pale brown with dark brown mottling along the hind margin. Body pale yellow in front, otherwise translucent white; becoming pink above and yellowish below in the last instar; anal plate pale brown marked with small black spots. Feeds July to September. When fully fed in late August or early September the larva leaves the fruit and spins a cocoon in a concealed place, such as under bark or in dead wood, where it overwinters before pupating in the spring. *Adult* emerges in June.

Cydia prunivorana
Larva: head reddish brown; prothoracic plate brown, translucent in front. Body whitish yellow, tinged with pink above; anal plate reddish black. Feeds July to September. When fully fed, the larva leaves the fruit and makes a cocoon under the bark where it spends the winter until it pupates in April. *Adult* emerges in May.

Pammene rhediella (fruit-mining tortrix)
Larva: head brown, darker in the early instars; prothoracic plate brown, sometimes yellowish brown with dark mottling. Body cream-white; anal plate greyish brown. Feeds June to August, primarily in the green fruits, but occasionally still present when the cherries are picked. The fully fed larva leaves the fruit and spins a cocoon, in bark or dead wood, in which it passes the winter before pupating in the spring. *Adult* emerges in May.

Pammene germmana
No description of the larva is available. *Adult* emerges in May.

27 Larva small and yellowish white; boring into shoot
family Yponomeutidae *Argyresthia pruniella*

Larva: head brownish black becoming paler with age. Body yellowish white. Feeds in March to May, initially on flowering shoots but later boring into the heart of the shoots and developing fruits, drawing them together in a light spinning. Pupates in the soil. *Adult* emerges in June.

–   Larva larger and pink and boring into shoot or
leaf stalk   family Tortricidae *Phtheochroa schreibersiana*

28 Larva exposed but under a silk web
family Yponomeutidae, subfamily Yponomeutinae 29

–   Larva exposed and not protected by silk while feeding,
although it may retreat into a roughly spun shelter
when not feeding   32

29 Larva solitary under a thin web; wriggles actively
when touched   *Swammerdamia pyrella*

Larva: head reddish brown; body yellowish, with a very fine red line along the middle of the back and stronger deep reddish lines towards the sides. Feeds under a light web on the upperside of the leaf in July, with a second generation in September. Pupates in a dense white cocoon amongst detritus on the ground, or occasionally on a twig on a tree. *Adult* emerges in May.

–   Larvae living together within a dense white web   30

30 Larva dark grey or greenish grey with rows of
black spots   31

–   Larva mottled reddish brown and greyish dark brown
Scythropia crataegella

Larva: head black with some whitish and brown lines. Body mottled reddish brown and greyish amber-coloured with the stripe along the back somewhat paler, especially on the first few segments where it is ochre-coloured. In September the larva feeds in a small transparent linear leafmine about 3 mm long; there are usually several larvae to a leaf. The larva then feeds externally before overwintering in a leafmine. In May and June larvae feeds in groups in a web. They pupate in the web. The pupa has an attractive shape and colour. *Adult* emerges in July.

PLATE 1

1. Raceme of *Prunus padus*

2. Flowering twig of *Prunus avium*

3. Webbing and damage on *Prunus padus* caused by the bird cherry ermine moth *Yponomeuta evonymella*

4. Damage on leaves of *Prunus padus* caused by *Phytodecta pallida*

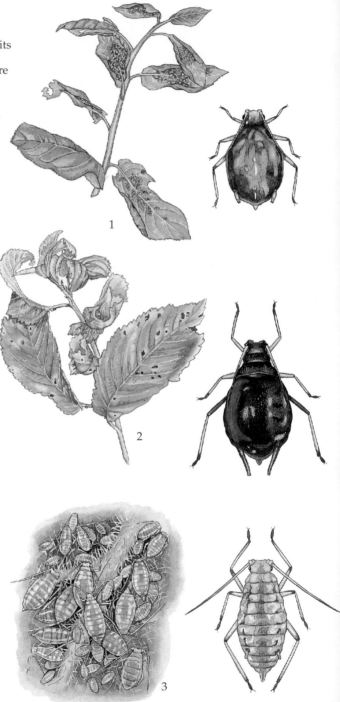

PLATE 2

1. Bird cherry aphid *Rhopalosiphum padi* and its pseudogalls on *Prunus padus*. Aphid colonies are visible on some leaves

2. The cherry blackfly *Myzus cerasi* and its leaf roll on *Prunus avium*

3. *Hyalopterus pruni* and a colony on cherry

    Individual aphids x 12

PLATE 3

1. *Apocheima pilosaria* (x 2.5)

2. *Erannis defoliaria* (x 3)

3. *Abraxas grossulariata* (x 3)

4. *Operophtera brumata* (x 3)

5. *Phalera bucephala* (x 1.5)

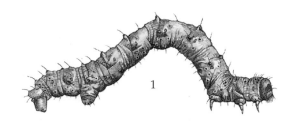

1

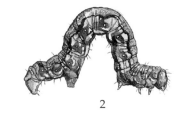

2

3

4

5

PLATE 4

1. *Anthocoris nemorum* (x 5)

2. *Pentatoma rufipes* (x 1.5)

3. *Phytodecta pallida* adult (x 5)

4. *Phytodecta pallida* larva (x 5)

5. *Coccinella 7-punctata* larva (x 5)

6. Hoverfly larva eating an aphid (x 4)

1

2

3

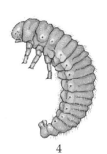

4

5

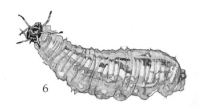

6

31 In webs, usually on *Prunus padus*; spots on the
sides of the back paired *Yponomeuta evonymella*

Larva: head black; body dark greenish grey with rows of black spots.
The black spots on the sides of the back are each divided into two.
Larvae feed briefly in early May by burrowing into buds and shoots,
occasionally causing drooping of the shoot, and then in a dense
communal web around shoots and leaves (pl. 1.3). They wriggle
vigorously when disturbed, often dropping to the ground on a silken
thread. They pupate in June or early July in a white cocoon within the
web. *Adult* emerges in July. This species is sometimes extremely
abundant in Scotland; may cause extensive defoliation of trees, but
without lasting consequences.

– In webs usually on *Prunus cerasifera*; spots on the sides
of the back single *Yponomeuta padella*

Larvae: head black; body dark leaden grey or greenish grey with rows
of black spots. The row of spots at the sides of the back conspicuous
and undivided. Briefly burrow into buds in early May, but then spin
around the shoots a dense web in which they live in groups. Wriggle
vigorously when disturbed and may drop to the ground on a silken
thread. Pupate in June in a flimsy grey cocoon within the communal
web. *Adults* emerge in June.

32 Larvae with characteristic arrangement of the false legs
(IV.8) and metred way of walking (hence they are called
'looper' caterpillars) family Geometridae (6 species) 33

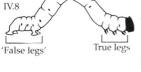

IV.8 'False legs' True legs

– Larvae with more typical lepidopteran arrangement of
false legs (IV.9) 38

33 Head black; body creamy white with black patches
Abraxas grossulariata

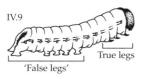

IV.9 'False legs' True legs

Larva: head black; body creamy white mottled with patches of black, a
reddish line joining the spiracles. Eggs laid in August produce larvae in
September which overwinter amongst detritus. In the spring, larva feeds
exposed until June when it produces an unconcealed black and
yellow pupa in a flimsy cocoon of only a few strands of silk. *Adults*
emerge in July.

– Head greenish or yellowish 34

34 Head pale green with a reddish spot on each side
Lomographa temerata

Larva: head pale green with a reddish spot on each side. Body green,
virtually unmarked in early stages, but ornamented with
brown-bordered reddish spots on back when full grown. Larva feeds
exposed in July to August, then pupates in a cocoon amongst detritus
in early September. *Adults* emerge in May.

– Head without a reddish spot on each side, although
sometimes dusted purplish 35

35 Body dark green with a series of purplish marks
along the back *Lomographa bimaculata*

Larva: head yellowish powdered with purplish. Body dark green with a
series of purplish marks along the back. Larva feeds exposed from June
to August, pupating in a dense cocoon amongst detritus in September.
Adults emerge in May.

– Body without such purplish marks 36

36 Body green with two white lines along it and a pale
    yellowish spiracular line on each side
                                    Operophtera brumata

Larva: head yellowish green; body green, rather short and stout, a stripe
of darker green along back, with two interrupted white lines each side
of it, a pale yellowish line joining the spiracles. Eggs laid in January to
February, hatching in late March. From April till June larva feeds at
night on the shoots of a wide range of trees and shrubs. By day the
larva stays concealed in a small chamber made by spinning several
leaves together. In June it pupates in a silk cocoon in the soil. *Adults*
emerge in November and December.

–   Body without white lines along it   37

37 Larva with only four yellowish lines along back and
    yellow stripe low down on side; matures in the middle
    of June   *Triphosa dubitata*

Larva: head yellowish green; body yellowish green usually with darker
green stripes and lines. Usually with four pale yellowish lines along the
back and a yellow stripe low down along the sides. Larvae hatch in
May and usually feed on *Rhamnus catharticus* and *Frangula alnus* but the
species has also been reared from *Prunus cerasus* (see table 1). Larva
makes a spinning of leaves in which it hides during the day. Pupates in
a cocoon in the soil in July. *Adults* emerge in August.

–   Larval colour as *dubitata* but with lines often greyish and
    with dark green line down middle of back (sometimes
    edges grey and incorporating a series of dark spots);
    matures at end of May   *Alsophila aescularia*

Larva: head pale green; body pale green with a rather darker line along
the back and yellowish lines along the sides. Larva feeds exposed in
May to June on a variety of deciduous trees; pupates in a cocoon in the
soil in early July. *Adults* emerge in March.

38 Larvae with complex many-pointed spikes (IV.10)
                                    family Nymphalidae (1 species)
          large tortoiseshell butterfly *Nymphalis polychloros*

Larva: up to 40 mm long. Black, with many white spots each with a
yellow forked spine; a single row of yellowish spots along the middle
of the back and two similar rows, one above the spiracles and the other
below. Larvae feed in groups from a conspicuous web on a wide range
of deciduous trees from May to July. Pupates exposed, suspended by
the caudal tip of the body in July. Pupa grey-black, strongly spiked on
the back and with golden metallic spots. *Adults* emerge in July. Possibly
extinct in Britain.

–   Larvae with simple hairs, although several may arise
    close together   39

39 Head blue-grey, body with 3 broad yellow stripes along it
                                    Diloba caeruleocephala

Larva: up to 40 mm long. Head blue-grey; body bluish grey with a small
black spot at base of each short hair, a broad interrupted yellow stripe
on the back and a similar broad yellow stripe on the side. Eggs laid in
November but not hatching until late April. Larva feeds externally
until August when it pupates in a cocoon amongst detritus. *Adults*
emerge in October.

–   Head brownish, body not as described above   40

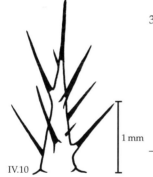

IV.10

1 mm

40　Body with divided hump on ring 11 and diamond pattern
along back　　　　　family Noctuidae *Allophyes oxyacanthae*
Larva: up to 50 mm long. Head reddish brown; body greyish or
brownish with dark grey and greenish mottling; a white-marked
divided hump on ring 11, three darker lines along the back with a
greyish diamond pattern between the outer ones. Eggs laid in
November but not hatching until April. Larva then feeds at night on a
variety of rosaceous trees until June when it spins a cocoon amongst
detritus. The larva is in diapause within the cocoon until July when it
pupates. *Adults* emerge in September.

–　Body without hump on ring 11 and with triangle pattern
along back　　　　　family Noctuidae *Xestia baja*
Larva: up to 40 mm long. Head light brown; body ochre-brown to
reddish brown; three yellowish lines along the back, the central one
edged blackish; the outer two with dark-bordered yellow triangular
marks between them. Eggs laid in August. From August to October,
larva feeds at night on herbaceous plants. Larva is in diapause from
November to March. After diapause, larva feeds at night on various
deciduous shrubs and trees until June when it pupates in a silk-lined
chamber in the soil. *Adults* emerge in July.

V Butterflies and moths Adult Lepidoptera

Many species of Lepidoptera take refuge in cherry trees during the day. The only species keyed out here are those whose larvae are known to feed on cherry. For species not described in this key refer to Heath & Emmet (1976–1996) (macromoths and micromoths); Bradley, Tremewan & Smith (1973, 1979) (Tortricidae); Goater (1986) (Pyralidae); Skinner (1984), or Brooks (1991) (Geometridae, Noctuidae & other large species); and Thomas & Lewington (1991) (butterflies).

1	Without functional wings (wings are absent or reduced to stubs)		2
–	With fully proportioned wings		3
2	Wing stubs present	*Operophtera brumata* female	
–	Wing stubs absent	*Alsophila aescularia* female	
3	Large species, wingspan greater than 25 mm		4
–	Small species, wingspan less than 25 mm		13
4	Body small in proportion to wings (V.1)		5
–	Body large in proportion to wings (V.2)		11

V.1

5   Antennae with a club at the end   butterflies

The only British buttefly species with larvae that feed on cherry is the large tortoiseshell butterfly *Nymphalis polychloros* (family Nymphalidae). This is very rare, and possibly extinct as a resident species in Britain, so check very carefully (in Thomas & Lewington, 1991) if you think you have found it. The small tortoiseshell butterfly *Aglais urticae* looks similar, and might rest in cherry trees, but its larvae feed on stinging nettles.

–   Antennae without a club at the end
                              family Geometridae (8 species) 6

V.2

6	Ground colour of wings white	7
–	Ground colour of wings brownish	9

7   Body yellow and black; wings heavily mottled with black blotches   *Abraxas grossulariata*
                                                    magpie moth

Wingspan 42–48 mm. The slender body is orange marked with black. Forewings white-cream mottled with large black spots which almost obscure a basal orange patch and a curved transverse orange band from halfway along rear edge of wing. *Adult* July to August; common in England, Wales and Ireland; less frequent in Scotland where it becomes a heathland species.

–   Body white   8

8   Forewings clean white with two triangular black-brown
     spots on front edge of wing   *Lomographa bimaculata*
                                             white-pinion spotted

Wingspan 27–30 mm. Forewings white with two small black triangular
marks on the leading edge. *Adult* May to June; widespread and not
uncommon in southern England, south Wales and the Lake District;
also recorded from Ireland.

–   Forewings silvery white irregularly suffused with
     grey-black towards the wing tip   *Lomographa temerata*
                                                 clouded silver

Wingspan 27–34 mm. Forewings white with silvery grey clouding
forming a band near to and parallel to the wingtip, a black smudge
near middle of wingtip and a small black spot in centre of forewing.
Adult May to June; widespread and not uncommon in England, Wales
and Ireland but scarce in southern Scotland.

9   Hindwings strongly notched between veins
                                               Triphosa dubitata
                                                       the tissue

Wingspan 38 - 48 mm. Forewings red-brown mixed with purple-brown
with vague bands of darker colour across them. Margins of hindwings
strongly scalloped. *Adult* August to May, hibernating in dense
vegetation from mid-Sepember to April; rather local throughout
England, Wales and Ireland.

–   Hindwings without notches between veins   10

10   Forewings grey-brown with two pale lines across them;
     hindwings pale with off-centre dark spot
                                          Alsophila aescularia male
                                                        march moth

Female brown and wingless. Wingspan of male 34–38 mm. Male
forewings brown with two wavy whitish lines across each wing. *Adults*
March to April; throughout the British Isles.

–   Forewings brown with many darker lines across them;
     no dark spot on hindwing   *Operophtera brumata* male
                                                        winter moth

Female brown with short wing stubs. Wingspan of male 28–33 mm.
Forewings yellowish brown with transverse wavy lines of darker
brown. Wingtips rather rounded. *Adults* occurring in late autumn and
throughout winter; common throughout most of the British Isles.

11   Hindwings whitish, forewing with bold central marks
                                         family Notodontidae (1 species)
                                               Diloba caeruleocephala
                                                      figure of eight

Wingspan 34–40 mm. Forewings silvery grey-fuscous to grey-brown
with two greenish cream dumb-bell-shaped patches behind the middle
of the leading edge of the wing. *Adults* October to early November.
Widespread in England and Wales but rather scarce in Scotland and
Ireland.

–   Hindwings brownish, forewings without contrasting
     marks in centre   family Noctuidae 12

12 Areas of forewing shiny emerald-green
Allophyes oxyacanthae
green brindled crescent

Wingspan 39–46 mm. Forewings dark brown, broadly bordered pale brown along edge of wingtip; a short black longitudinal line from base of wing and a short curved white line from near the middle of rear edge of wing. Whole forewing with silvery green reflections. *Adults* September to October. Widespread throughout the British Isles; scarce in northern Scotland although common elsewhere.

– Areas of forewing without green streaks; uniform brown with black spot on vein that forms front margin of wing, before the tip *Xestia baja* dotted clay

Wingspan 38–44 mm. Forewings various shades of brown with a short black oblique dash near the tip of the leading edge. *Adult* July to August. Widespread and common throughout the British Isles.

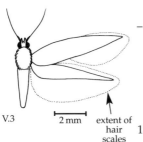

V.3   2 mm   extent of hair scales

13 Forewings with pointed ends (V.3)   14

– Forewings square-ended or nearly so (V.4)   30

14 Base of antennae with scaly projection (or eye-cap) that conceals the eye when viewed from above (V.5)   15

– Base of antennae without eye-caps   18

15 Forewings with a pale band across it; wingspan 4–6 mm
     16

– Forewings not like this; wingspan greater than 6 mm   17

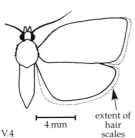

V.4   4 mm   extent of hair scales

16 Head covered with orange scales (eye-caps white)
Ectoedemia atricollis

Wingspan 5–6 mm. Head orange-brown, collar black and eye-caps whitish. Forewings brownish black with a narrow silver band across the middle. *Adult* in June. Widespread and common throughout England and Wales, just getting into southern Scotland.

– Head covered with black scales (eye-caps white)
Stigmella prunetorum

Wingspan 4–5 mm. Head black but collar and eye-caps whitish. Forewing with a broad silvery band at half way, a transverse band before this silvery band and the area beyond it purplish brown. Basal third shining golden. *Adult* May and again in August. Scarce and local in southern and western England.

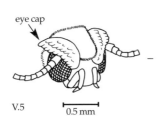

eye cap

V.5   0.5 mm

17 Eye-caps small and dark, forewings bronzy-yellow-brown and without markings *Tischeria gaunacella*

Wingspan 6–8 mm. Head and thorax leaden metallic. Forewings brownish grey. *Adult* May to June. Taken last century at two places in Essex but not seen since.

– Eye-caps well developed and white; forewings silvery white or drab grey with dark streaks and other marks
Lyonetia clerkella
apple leaf miner

Wingspan 8–9 mm. Forewings silvery white to drab grey with streaky dark markings at tip. Wings very narrow. *Adult* June, August and sometimes in October. The October moths overwinter in dense cover until April. Common throughout the British Isles.

18 Forewings golden-brown with four white triangular marks on front edge of wing and three on the back edge 19

– Forewing unmarked or marked otherwise 20

19 Thorax golden-brown with white line along the middle
Phyllonoryter cerasicolella

Wingspan 7–8 mm. Thorax orange-brown with white front edge and central line. Forewings bright golden brown with four whitish triangular marks along the front edge and three along the rear edge; each bordered blackish along the basal edge. *Adult* May and again in August. Widespread and locally common throughout the southern half of England and in south Wales.

– Thorax golden-brown but without central line
Phyllonoryter sorbi

Wingspan 7.5–8.5 mm. Thorax golden brown with white along front edge. Forewings shining golden brown with four whitish triangular marks along the front edge and three along the rear edge; each bordered blackish along basal edge. *Adult* April to May and again in August. Common throughout the British Isles.

20 Fringe on rear edge of hindwing at least three times width of hindwing in middle 21

– Fringe on rear edge at most twice width of hindwing in middle 25

21 Forewings basically white with speckling 22

– Forewings with at most the leading edge white 23

22 Forewings white, heavily sprinkled with dark scales and distinct dark spot near centre of wing
Coleophora hemerobiella

Wingspan 12–15 mm. Forewings white speckled with orange-brown and with a dark spot in middle of wing two-thirds of the way from the base. *Adult* July. Widespread in the southern half of England.

– Forewings white, lightly sprinkled with dark scales; no central spot *Coleophora anatipennella*
pistol case-bearer

Wingspan 12.5–17 mm. Forewings white, lightly speckled with orange-brown. *Adult* July. Throughout England and Wales but very local in Scotland and Ireland.

23 Leading edge of forewing broadly whitish or yellowish 24

– Forewings without white leading edge, entirely plain greyish brown *Coleophora coracipennella*
C. prunifoliae
C. spinella

These species can only be safely separated by inspection of the genitalia (see Emmet and others, 1996). All three species have wingspan 9–12 mm and plain orange-brown forewings.

24 Forewings without dark-tipped scales or with a few
near the wing-tip   *Coleophora adjectella*

Wingspan 9–10 mm. Forewings ochreous rust-coloured tinged
brownish but pale ochreous yellowish along basal half of leading edge.
Adult June to August. Widespread in south-east and southern England.

– Forewings with many dark-tipped scales in the third of
the wing nearest the wing-tip   *Coleophora trigeminella*

Wingspan 9–10 mm. Forewings ochreous rust-coloured tinged with
grey but creamy white along basal three-quarters of leading edge. *Adult*
June. Very local in south-east England and parts of the Midlands.

25 Forewings white (sometimes greyish) with numerous
black spots   26
– Forewings not spotted   27

26 Forewings with five or six rows of small dots, more than
eight in row along rear edge   *Yponomeuta evonymella*

Wingspan 19–25 mm. Head white; thorax and forewings white with
many black spots. Forewings with five or six rows of black spots with
the row along the rear edge of the wing with more than eight spots.
Adult July to August. Widespread and common throughout Britain.

– Forewings with three or four rows of small dots, fewer
than eight in row along rear edge   *Yponomeuta padella*

Wingspan 19–22 mm. Head white; thorax and forewings pale greyish
white or grey with many black spots. Forewings with three or four
rows of black spots with the row along the rear edge of the wing with
fewer than eight spots. *Adult* June to August; widespread and often
common, especially in southern England.

27 Head white, thorax grey   28
– Head white, thorax white   29

28 Forewings predominantly grey with wingtips coppery
Swammerdamia pyrella

Wingspan 10–13 mm. Head yellowish white, thorax grey and forewings
greyish fuscous with a few paler and darker areas. Tip of wing with
strong coppery reflections. *Adult* May and again in August. Common
throughout Britain as far north as central Scotland.

– Forewings predominantly grey without coppery tips
S. caesiella

Wingspan 9–13 mm. Head whitish, thorax grey and forewings greyish
fuscous with a few paler and darker areas. *Adult* May and June and
again in August. Common throughout the British Isles.

29 Forewing white with two brownish bands across it
and brown speckling *Scythropia crataegella*

Wingspan 11–15 mm. Head and thorax white but abdomen grey;
forewings white each with a pair of broad transverse orange-brown
bands and speckles of the same colour. *Adult* July. Widespread and
often common in southern England.

– Forewings yellowish and brownish with rear edge
broadly white except for dark square spot halfway along
Argyresthia pruniella
cherry fruit moth

Wingspan 10–12 mm. Head and thorax white, but abdomen
grey-brown. Forewings orange-brown at base merging to dark brown
at wing tip. A strong white band along rear edge of wing interrupted
by a dark brown square mark about the middle. *Adult* July. Common
throughout the British Isles except for northern Scotland.

30 Tip of forewing forming an acute angle 31
– Tip of forewing nearly a right angle 41

31 Basal third of forewing pale yellowish, remainder
predominantly yellow-brown and black
Phtheochroa schreibersiana

Wingspan 11–15 mm. Forewings dark brown mixed black and bluish;
basal third ochreous and orangey-white and another vague spot
halfway along leading edge similarly coloured. *Adult* May to June. Very
local in southern England.

– Basal third not distinctly paler than rest 32

32 End part of forewing predominantly whitish; remainder
patterned dark grey-black 33
– End part of forewing not whitish 35

33 End two-thirds of forewing predominantly whitish
although overlaid with some chestnut markings
Epiblema trimaculana

Wingspan 15–17 mm. Forewings with basal third blackish brown,
remainder white clouded with grey, except for the area of the wing tip
and a partial band at two-thirds of the way along rear edge, which are
predominantly chestnut-coloured.

– Only end third of forewing whitish 34

34 Extreme tip of forewing blackish *Hedya pruniana*

Wingspan 15–18 mm. Basal two-thirds of forewing patterned black,
brown and greyish; apical third whitish with some clouding and
extreme wing tip blackish. *Adult* June to July. Throughout the British
Isles.

– Extreme tip of forewing sometimes pale greyish,
but usually creamy white *Hedya dimidioalba*

Wingspan 15–21 mm. Basal two-thirds of forewing patterned black,
brown and greyish; apical third whitish clouded with grey. *Adult* June
to July. Throughout the British Isles as far north as mid Scotland.

35 Forewings brownish with a streak of blackish extending
from the wing tip to moth's body; leading edge of wing
at tip without paler angular marking *Acleris umbrana*

Wingspan 18–22 mm. Forewings brownish with velvet black streak
from wing tip to base and a prominant tuft of scales near the middle.
Adults August to April, hibernating over winter. Very local in England
as far north as Northumberland.

– Colour and markings not like this; leading edge of
forewing always with pale angular markings near tip 36

36 Forewing with some orangey chestnut colouration 37
– Forewing primarily brownish or blackish 38

37 Only outer third of forewing orangey chestnut,
remainder dark brownish *Pammene rhediella*
fruit-mining tortrix

Wingspan 9–12 mm. Forewings dark brown with apical third chestnut
orange streaked with grey. *Adult* May to June. Throughout the British
Isles.

– Whole of forewing patterned with lines of orangey
chestnut, black and grey on a ground of dark brownish
Enarmonia formosana
cherry bark tortrix

Wingspan 15–19 mm. Forewings dark brown finely patterned with
orange, black and grey lines. *Adult* June to August. Throughout the
British Isles as far north as central Scotland.

38 Pale angular marks on leading edge of forewing
extending along whole length 39
– Pale angular marks on leading edge of forewing only
distinct near wing tip 40

39 Forewings velvety black-brown with pale angular
markings contrasting creamy. Smaller species
10–14 mm *Pammene germmana*

Wingspan 10–14 mm. Forewing uniform dark brown with contrasting
cream angular marks along leading edge. *Adult* May to June. Local in
England.

– Forewings patterned light and dark brown, very finely
speckled pale yellowish. Larger species 14–16 mm
Epinotia signatana

Wingspan 14–16 mm. Forewings cream-white mixed with fuscous, with
a vague brown basal patch and an indistinct oblique brown band from
the middle of the leading edge. A silvery grey ring near tip of wing.
Adult June to July. Local and uncommon throughout England and
Wales.

40 Forewings fuscous brown patterned with unclear
whitish and grey markings *Cydia funebrana*
plum fruit moth
red plum maggot

Wingspan 10–15 mm. Forewings fuscous brown obscurely patterned
darker and with purplish lustre; area of wing tip marked with grey.
Adult June to July. Locally common in Wales and southern England as
far north as Lancashire and Yorkshire.

– Forewings ochreous brown patterned similarly
Cydia prunivorana

Wingspan 13–14 mm. Forewings ochreous brown obscurely patterned
darker; area of wing tip marked with grey. *Adult* May to June.
Restricted to south-east England but widespread in Europe.

41 Forewings with a distinct broad sloping band
extending the full width of the wing 42
– Forewing band not like this or absent 43

42 Forewings with inner edge of broad transverse band
almost parallel to edge of basal patch; wings with a
strong net-like pattern *Pandemis corylana*

Wingspan 18–24 mm. Forewings ochreous yellow, with a net-like
pattern of ochreous brown or dark brown with a basal patch and a
broad transverse band of darker ground colour. Oblique outer margin
of basal patch parallel to inner margin of transverse band. *Adult* July to
September. Throughout the British Isles but scarcer in Scotland.

– Forewings with inner edge of broad transverse band and
edge of basal patch diverging from front to back; wing
with little or no net-like patterning *Pandemis cerasana*

Wingspan 16–24 mm. Forewings yellowish brown with a basal patch
and a broad transverse band of darker brown. Outer margin of basal
patch slightly oblique while inner margin of transverse band strongly
oblique. *Adult* June to August. Common throughout the British Isles.

43 Forewings light grey-brown marked with dark brown
suffused with ochreous. In the male the base of the
leading edge is folded backward *Adoxophyes orana*

Wingspan 15–22 mm. Forewings greyish brown with a series of
obliquely transverse bands or lines of darker brown. *Adult* June and
again August to October. First found in Britain in 1950, now
widespread in orchards throughout southern England.

– Forewings with ground colour always yellowish or
reddish, never greyish 44

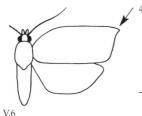

V.6

44 Tip of forewings sickle-shaped (V.6) *Acleris rhombana*

Wingspan 13–19 mm. Tip of forewing sickle-shaped. Forewings vary
from ochreous yellow to almost reddish brown, usually marked in a
net-like pattern of darker brownish. Markings variable but usually with
an oblique short band of dark brownish halfway along front edge of
forewing. *Adult* August to October. Common throughout British Isles.

– Tip of forewings not sickle-shaped (V.7) *Acleris notana*

Wingspan 15–19 mm. Forewings vary from pale ochreous to reddish
ochreous, sometimes speckled brown or black. Markings variable but
often with a brownish or blackish triangular patch halfway along the
leading edge. *Adults* July and October to April in England and Wales,
but August to May only in Scotland, hibernating overwinter in dense
vegetation. Widespread throughout British Isles.

V.7

VI Sawfly larvae Hymenoptera

See Cameron (1882–1892) for further information.

1 Larva with a pair of rod-like projections below the anus
and feeding within a rolled or folded leaf
Pamphilius sylvaticus

The solitary green larva with the characteristic projections below the
anus spends the day concealed in a folded leaf edge, venturing out at
night to feed on the nearby leaf (June to July). When fully fed, it
descends and pupates underground. Common throughout Britain on
various rosaceous trees.

– Larva without projections below the anus and feeding
in an exposed position 2

2 Larva slug-like, blackish and slimy *Caliroa cerasi*

Larvae can reach pest proportions on both cherry and pear; other
rosaceous trees are also attacked. The larva is whitish when it hatches
from the egg, but soon becomes dark olive-brown and slug-like. Its
appearance has earned it the name 'Cherry Slug-worm' because the
thoracic segments are enlarged into a big shiny hump. The rest of the
body tapers to the tail. The shininess is due to the production of slime.
The larva skeletonises the leaves, leaving one cuticle intact. In the final
larval instar it reverts to the more conventional sawfly larva shape and
becomes orange. It descends to the ground and pupates in a stout
cocoon in the soil. There are several broods per year and the species
overwinters as a mature larva within the subterranean cocoon.
Common locally throughout Britain, usually in at least two broods.

– Larva not like this 3

3 Larvae solitary 4

– Larvae living in groups in a communal web
Neurotoma saltuum

The yellow or orange larvae have black heads and live communally in
webs on various rosaceous trees, most commonly on pear. The webs are
most frequent in June and July. Very local in the southern half of
England.

4 Larva with orange tint *Caliroa cerasi* (final instar)

– Larva not tinted orange *Pristiphora retusa*

VII Adult sawflies Hymenoptera

For further identification see Benson (1951, 1952, 1958) &
Wright (1989).

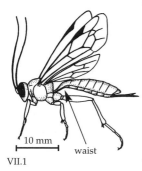

VII.1

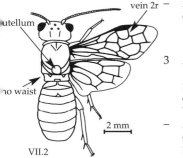

VII.2

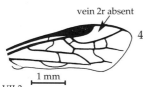

VII.3

1   Wasp-like, with a well-developed waist between
thorax and abdomen (VII.1)

Not a sawfly

Probably either (a) a wasp or bee collecting honeydew released by
aphids, or (b) a parasitic wasp hunting for a host. Neither of these is
included in the present key.

–   Without a distinct waist between thorax and abdomen
(VII.2)   A sawfly 2

2   Hind edge of pronotum (top of segment bearing the first
pair of legs) almost straight; on forewing, vein 2r is
present (VII.2)   family Pamphiliidae (2 species) 3

–   Hind edge of pronotum distinctly curved (not straight);
on forewing, vein 2r is absent (VII.3)

family Tenthredinidae (2 species) 4

3   Abdomen black except at the tip; scutellum (VII.2) yellow
Pamphilius sylvaticus

Body length 8.5–10.5 mm. Adult with head, thorax and abdomen,
except for tip, black. Part of face and scutellum yellow. Wings yellowish
with brown veins becoming yellow at base. May to June.

–   Abdomen orange to blackish orange with pale flecks
on the sides; scutellum black   *Neurotoma saltuum*

Body length 8.5–10.5 mm. Adult with head, thorax and base of
abdomen black; legs and part of face yellow; abdomen orange, to a
greater or lesser extent, with yellow flecks on the sides on most
segments; wings with a smoky band across the middle. May to July.

4   Small species, 3–4 mm long; mainly black, partially
yellow along sides; hindlegs predominantly pale
Pristophora retusa

Body length 3–4 mm. Adult mainly black but with yellow on the
mouthparts, edge of pronotum, tegulae (small plates that cap the wing
bases) and the sides of some of the abdominal segments. Hind legs
predominantly pale. June. In Britain known only from Inverness-shire.
Rare in Europe.

–   Slightly larger species, 4–6 mm long. Entirely black
including the hindlegs   *Caliroa cerasi*

Body length 4–6 mm. Adult totally black and parthenogenetic (males
are very rare). May to September.

VIII Fly larvae Diptera

For further identification of reared adult flies see Colyer & Hammond (1968) and Unwin (1981) (general); Barnes (1948, 1951) (Cecidomyiidae); Spencer (1972) (Agromyzidae); White (1988) (Tephritidae). Smith (1989) deals with immature stages of flies.

anchor process

VIII.1

0.5 mm

1 Very small spindle-shaped maggots with chitinous 'anchor process' on underside of first thoracic segment (VIII.1) (use a lens or microscope to see this)
family Cecidomyiidae (4 species) 2

Adult cecidomyiids have long thin antennae and slender bodies (VIII.2) and cannot be identified to species without making microscope slides.

– Maggot without anchor process 5

2 Larva white or whitish 3

– Larva reddish yellow or yellowish orange 4

3 Larvae live in groups, enclosed in a galled shoot
Dasineura tortrix

Groups of white larvae enclosed within a gall formed by the terminal leaves of the shoot becoming massed together because internodes stop growing. The leaves involved become rolled within each other with their margins loosely rolled upwards.

– Larvae solitary, associated with aphids
Aphidoletes aphidimyza

Larvae predatory on aphids, especially *Myzus cerasi*

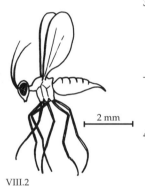

VIII.2

2 mm

4 Larva reddish yellow, solitary and enclosed in a bud gall
Asphondylia prunorum

Solitary reddish yellow larva in a large cavity within an oval bud gall 4–5 mm high. The bud gall is at first green becoming yellowish or reddish, with a brown tip and brown scales surrounding the base. Larva always seems to be associated with a fungus within the gall.

– Larvae yellowish orange, living in groups in pouch-like gall on leaf vein *Putoniella pruni*

Groups of yellowish orange larvae within a leaf gall. The midrib or another vein swells into a pouch-like structure on the underside of the leaf with a narrow opening above.

5 Three options:

– Larva elongate, living in a tunnel in living wood
family Agromyzidae *Phytobia cerasiferae*

The slender, elongate white larva, which may be up to 15 mm long, bores in the cambium of the twigs and trunk throughout winter. In early spring it leaves the mine and falls to the ground where it pupates. The larval damage reduces the commercial value of the timber. The adult is a small stout short-bodied grey fly (wing length 3 mm) with clear wings.

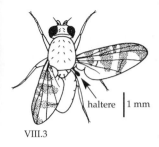

haltere   1 mm

VIII.3

–   Larva short and stubby, living in fruit
family Tephritidae *Rhagoletis cerasi*
cherry fly

The short, stubby white larva feeds in the soft tissue of the cherry. Not native in Britain but sometimes introduced in imported cherries. It has not as yet become established in the wild. The adult, known as the 'cherry fly', is a small, dark brown or black fly with a yellow stripe on the side and yellow scutellum and lower legs. The characteristically patterned wings have three dusky bands across them and a band around the tip (VIII.3).

–   Larva variously coloured, tapered towards the head and living on surface of leaves and stems
hoverfly, family Syrphidae
Not covered here; see Rotheray (1993) for larvae or Stubbs & Falk (1983) for adults reared from them.

IX Adult beetles Coleoptera

For further identification see Joy (1932) and Hodge & Jones (1995) (general); Levey (1977) (Buprestidae); Kirk-Spriggs (1996) (Nitidulidae); Morris (1990) (weevils); Duffy (1953) (Scolytidae) and Duffy (1952) (Cerambycidae).

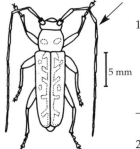

IX.1

1   Antennae long and thread-like; all antennal segments of similar length (IX.1)
family Cerambycidae *Saperda scalaris*

Length 14–16 mm. Antennae thread-like and longer than length of body. Body black; head, antennae, legs and underside clothed with bright yellow-green hairs. Elytra patterned with same pubescence. Larva bores into dead wood. Widespread but rare throughout England.

–   Antennae not like this   2

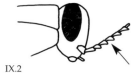

IX.2

2   Antennae comb-like, with all segments of similar size (IX.2)   family Buprestidae (2 species) 3
–   Antennae clubbed, broadened at or near tip   4

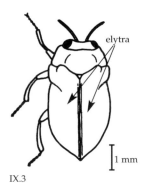

IX.3

3   Elytra (hardened forewings, IX.3) with four wavy white bands of fine hairs across them; body 3–3.5 mm long
Trachys minutus

Body oblong, tapered at rear; entirely black except for four wavy bands of whitish hairs across the elytra. Larva occurs on willows *Salix* species and hornbeam *Carpinus betulus* more commonly than on cherry. Adult hibernates and then can be beaten from the larval foodplant in mid May to late July. Larva cream-coloured, soft and legless with a well-developed brown chitinous head. Feeds within a leaf. Leaf-mine starts with a small, shiny fleck at the leaf-margin and progresses to a large pear-shaped blotch in the upper surface of the leaf. It is greenish when fresh but soon goes brownish. Pupation takes place naked (not in a cocoon) within the mine. Local but widely scattered in southern England.

–   Elytra not like this; body 4.5–7 mm long   *Athaxia nitidula*

Body oblong, tapered at rear; entirely golden green except for black eyes. Adults can be beaten from the larval foodplant in mid May to late July but also frequent the flowers of roses, buttercups and hawthorn. The soft, legless larva is cream-coloured with a well-developed brownish chitinous head, and lives beneath the bark. The larva is broadest at the first thoracic segment, and the body tapers towards the tail. On the very western edge of its range in Britain; known only from the New Forest. No recent records.

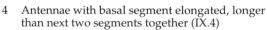

4 Antennae with basal segment elongated, longer
than next two segments together (IX.4) 5
– Antennae with basal segment shorter than next two
segments together 13

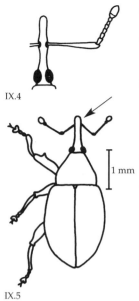

IX.4

5 Head prolonged into a rostrum (snout-like projection)
(IX.5) Weevils family Curculionidae 6
– Head without distinct rostrum; antennae of three to
seven segments in total family Scolytidae 12

6 Rostrum clearly visible from above; body length
more than 2 mm 7
– Rostrum not visible from above; very small species,
length 1.5–2 mm *Ramphus pulicarius*

Rostrum present but always bent under the body; antennae originate
near base of rostrum. Body blackish and without hairs. Antennae with
basal portion reddish. Distributed throughout the British Isles.

IX.5

7 Blackish, long and narrow genus *Magdalis* (2 species) 8
– Reddish brown to dark brown 9

8 Rostrum as long as head plus thorax *Magdalis cerasi*

Length from base of rostrum to tip of abdomen 2.5–3.5 mm. Rostrum as
long as head and thorax together; antennae originate two-thirds of the
way to the tip. Body dull blackish. In England as far north as
Lancashire.

– Rostrum clearly shorter than head plus thorax
Magdalis ruficornis

Length from base of rostrum to tip of abdomen 2.5–3.5 mm. Rostrum
only three-quarters as long as head plus thorax; antennae originate
two-thirds of the way to tip of rostrum. Body dull blackish but basal
portion of antennae reddish. Throughout Britain although rarer in the
north.

9 Shape short and broad (IX.6); rostrum clearly longer
than head plus thorax 10
– Shape more ovoid (IX.7); rostrum similar in length to
head plus thorax *Rhynchaenus testaceus*

Length 3–3.5 mm. Rostrum as long as head plus thorax; antennae
originate half way along rostrum. Body reddish brown, paler towards
the tip but with a darkish mark half way along the outer edge of the
elytron. Widespread throughout British Isles.

IX.6

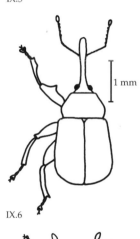

IX.7

10   Larger species, body length more than 4.0 mm; femora of
     first pair of legs with distinct double tooth

                                                   Furcipus rectirostris

Body length (including rostrum) 5–6 mm. Rostrum slightly longer than
head plus thorax. Head, thorax and elytra reddish brown. Elytra
patterned with a yellowish band across them two-thirds of the distance
from the base, a white line down the midline where the two elytra abut
and a shorter longitudinal white line each side of this. Eggs laid singly
in the ovary of a dying flower. The larvae feed on the developing kernel
of the cherry and pupate within the wood shell. The adults emerge
from the ripe cherries and feed on the leaves before hibernating. Very
local, mostly in northern England, on *Prunus padus* (Read, 1981).

–   Smaller species, body length less than 3.5 mm; femora of
     first pair of legs with single tooth (IX.6)

                                     genus *Anthonomus* (2 species) 11

11   Pale bar across elytra two-thirds of the way back
     from the head   *Anthonomus bituberculatus*

Length 2–3 mm. Rostrum nearly 1.5 times length of head plus thorax;
antennae originate three quarters of the way to tip of rostrum. Body
usually reddish brown, sometimes dark red-brown; a pale bar across
elytra two-thirds of the way from the head; front and rear borders of
elytra are dark. Eggs laid into a bud, which is not retarded. Larva feeds
wwithin the developing flower preventing it from opening. Adult
hibernates over winter under bark. Widespread in southern England.

–   Elytra with only a slight trace of whitish colour

                                               Anthonomus humeralis

Length 2.5–3 mm. Body dark brown with only slight traces of whitish
on elytra. Larva probably develops in the flower as in the previous
species. Very local in southern England.

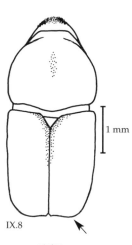

IX.8

1 mm

12   Rear edge of elytra squarish (IX.8)   *Scolytus mali*

Length 3–4.5 mm. Black, with antennae, legs and elytra rusty brown.
From above, somewhat oblong in shape. Female makes a gallery up to
12 cm long, longitudinally in the sapwood of a wide variety of dead
rosaceous trees. She lays eggs at intervals along this gallery. When the
larvae hatch they make galleries which are narrow at first and then
widen gradually, perpendicular to the maternal gallery. Pupation is
either in the bark or the sapwood. Rather local in the southern half of
England.

–   Rear edge of elytra ovoid (IX.9)   *Scolytus rugulosus*

Length 1.5–3 mm. Pitchy-black, elytra usually lighter with rear third
reddish brown. Antennae and legs rusty brown. Appears ovoid from
above. Female makes a short gallery 1–4 cm long, longitudinally in the
sapwood of a wide variety of dead rosaceous trees. The larval galleries
are at right angles to the maternal gallery and are often dense and
intermixed. Pupation is deep in the wood. Rather local in the southern
half of England.

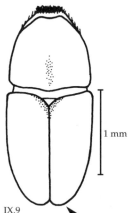

IX.9

1 mm

13   Antennae with basal segment very short, shorter than
     next segment   family Attelabidae (5 species) 14

–   Antennae with basal segment slightly longer than
     next segment   18

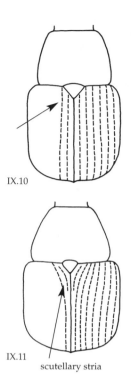

IX.10

IX.11

scutellary stria

14 Elytra without a scutellary stria (IX.10)   15
 – Elytra with a distinct scutellary stria (IX.11)   17

15 Elytra, at least in front, thickly and irregularly pitted; length 6–9 mm   16
 – Elytra regularly pitted; length 2.5–3 mm
                                                 Rhynchites caeruleus

Body shining deep blue, clothed with long upright yellow-brown hairs. Rostrum as long as head (male) or as long as thorax (pronotum) (female). Recorded from most species of rosaceous trees. Egg is laid into a young shoot which the female then partially severs. Larva develops in the dying shoot initially on the tree but also in the fallen shoot on the ground. Occasionally a pest in orchards where it has earned the name 'Twig-cutting weevil'. Widespread and sometimes locally common in Britain as far north as Durham.

16 Upper surface of rostrum metallic   *Rhynchites auratus*

Length 8–9 mm. Body greenish or golden-coppery clothed with long white hairs; elytra edged with red. Pronotum of male armed with a sharp, forward-directed spine at the side in front. Rostrum as long as thorax. Larva lives in kernels of the stones of *Prunus* species. There are old records from (mostly) southern England but the species is now thought to be extinct in Britain.

 – Upper surface of rostrum black   *Rhynchites bacchus*

Length 6–8 mm. Body crimson or golden-coppery or purple and clothed in long yellow-brown hairs. Rostrum as long as head and pronotum combined. Larvae in the fruit of various rosaceous trees and shrubs. There are old records for southern England but the species has not been seen this century and is presumed extinct.

17 Head and pronotum metallic-coppery-bronze or dark brown and elytra a contrasting brick-red, reddish brown or orange-brown   *Rhynchites aequatus*

Length 2–4 mm. Body dark brown with slight golden reflection. Elytra reddish except at midline in front where blackish. Rostrum as long as head and pronotum combined. Larvae in the fruits of wide range of rosaceous trees and shrubs. Sometimes a pest of apple, hence the name 'Apple Fruit Rhynchites'. Widely distributed and often common throughout England and into Scotland.

 – Upper surface entirely blue or dark blue
                                              Rhynchites pauxillus

Length 2–2.5 mm. Body shining deep blue, less evidently clothed in hairs than *R. caeruleus*. Widespread but rare throughout England and into southern Scotland. Larvae sever the leaf stalk of various rosaceous trees.

18 Antennae with a large spherical club composed of three roughly equal segments; basal segment swollen
              family Nitidulidae (1 species) *Meligethes coracinus*

Length 2 mm. Body oblong-oval, convex, rather dull black, sometimes with a very slight greenish or bronze reflection. Very rare in southern England.

 – Antennae with a long club of five or six segments; basal segment not swollen
              family Chrysomelidae (1 species) *Phytodecta pallida*

Length 5–7 mm. Body oblong-oval, convex, brown, sometimes with a suggestion of darker markings on the elytra. *Adult* May to June. Throughout the British Isles.

X Adult heteropteran bugs Hemiptera: Heteroptera

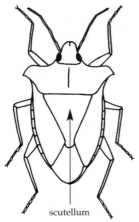

X.1

scutellum

When viewed from above, the hind part of a heteropteran bug seems to show two triangles: there is a backwards-pointing triangle formed by the scutellum (X.1), and behind that a forward-pointing one formed from the overlapping membranous tips of the forewings. Many species of heteropteran bugs settle on cherry accidentally. If feeding is not observed the extended keys of Southwood & Leston (1959) should be consulted.

1 Antennae with five segments
family Pentatomidae (1 species) *Pentatoma rufipes* forest bug

Length 12–15 mm. Flattened, shield-shaped bug with pronotum extended in lateral square-ended processes. Body bronzy, speckled with black pitted marks. An orange spot in the middle of the back two-thirds down towards rear. Behind this is a square area the front half of which is black and the back half of which consists of transparent wing membranes. Body edged black and orange towards the rear, legs orange-brown. Eggs laid in August, hatch in August to September. The half-grown larva hibernates over winter under bark and begins to feed again in April to early May. Feed by sucking juices from trees and also caterpillars. *Adults* July to September, thus sometimes common on cherry orchards at picking time. Common in woods and orchards throughout Britain extending well into Scotland.

– Antennae with four segments
family Miridae (4 species) 2

2 Body predominantly dark brown with creamy markings 3
– Body predominantly greenish 4

3 Legs hairy; dark brown mottled with cream
Phytocoris dimidiatus

Length 6–8 mm. Body elongate-oval, dark brown with a few small creamy markings; head creamy with dark brown eyes; legs cream-coloured heavily mottled dark brown. Whole insect covered with conspicuous dark brown hairs. Eggs laid on trees in the autumn, overwinter, and hatch in May to June. Probably mostly predatory on small insects. *Adults* June to November. Widely distributed in the British Isles on a variety of deciduous trees.

– Legs smooth; cream to pale brownish
Globiceps flavomaculatus

Length 6–8 mm. A narrow elongate oval bug of predominantly dark brown body colouration and with yellow-brown legs. A pale creamy blotch on each side of the body at one-third back, and another similar one at two-thirds back. Eggs laid in cracks in the bark of a variety of trees and shrubs. Larvae and adults feed by sucking juices from plants and from other animals. *Adults* June to September. Female usually short-winged.

4   Body elongate oval; antennae as long as body
                                        Lygocoris pabulinus
                                  common green capsid

Length 5–7 mm. An oval, plain green bug that has two generations per
year each using both a woody and a herbaceous host plant. Eggs laid in
September on young twigs of woody host plant hatch into green larvae
the following April. These feed on the woody host for a few weeks and
then migrate on to low herbs such as nettles, thistles, dandelions or
willowherb where they become adult in June. Eggs are laid by these
adults in June to July in the stems and leaf stalks of the herbaceous host
plant. The second generation adults occur in August to September and
return to the woody host to lay eggs. Widespread and abundant
throughout the British Isles.

–   Body parallel-sided; antennae distinctly shorter
    than body   *Orthotylus marginalis*
                                  dark green apple-capsid

Length 5–7 mm. A narrow elongate oval bug with green body and legs,
a collar of yellow just behind head and yellow midline spot a little back
from this. Eggs laid in the young wood near the buds and overwinter,
hatching in April to May. Feeds by sucking plant juices but also known
to prey upon red spider mites, aphids and other small insects. *Adults*
mid June to late August. Widespread throughout the British Isles.

5 Techniques

In this chapter we describe some methods that are likely to be useful in studies of the insect fauna of cherry trees.

Collecting and looking for insects

Basically, only three types of collecting equipment are needed when working with cherry tree dwelling insects: a beating tray and stick, a pooter or insect aspirator, and a good collection of different sized tubes, cartons, boxes, jars and bottles. A quick hand and eye are also useful! A small hand lens is also a great help, as some of the insects, such as aphid nymphs, are small.

Searching by hand and eye is perhaps the most sensible way to start, especially if you are looking for a specific type of insect or damage effect. Galls are easily visible and most leaf-feeding insects, such as leaf miners and aphids, are not greatly disturbed if you are careful when you examine the leaf. Remember to look at both the underside and the top. Different insects have different preferences. Aphids for example are generally found on the underside of leaves. When censusing a population, this non-destructive sampling method is ideal as the colony or individual insect is left untouched and the foliage of the plant unharmed. It is then possible to come back the next day or week, to the same leaves or branches (if they have been marked beforehand) or to the same tree and see what has happened to the insects in the meantime.

For a one-off sampling exercise, such as counting the insects on a particular tree at one fixed point in the season, a beating tray can be used. A beating tray is basically a folding wooden or metal frame supporting a white cotton cloth cover. A pale-coloured umbrella held upside down is an acceptable substitute. The beating tray is held under a selected branch, which is then hit a fixed number of times with the beating stick. The insects that fall onto the tray can then be collected using a pooter or fine pair of forceps, and put into suitable containers. They can then be identified and counted back in the laboratory.

It is sometimes worth collecting branches or leaves and placing them in a polythene bag, taking them back to the laboratory and either waiting to see what comes out or sorting through the material carefully with a hand lens. This destructive sampling method will pick up the smaller insects that inhabit stems and leaves, and can yield information on leaf size and weight which may correlate with insect numbers.

When collecting insects it is a good idea to note what time of day it is and the weather, particularly the temperature. Some insects feed at particular times of day and are more noticeable at those times, and temperature has a marked effect on insect activity. Sampling a tree or group of trees throughout the year should reveal how the insects associated with it change as the seasons progress.

Killing and preserving insects

Not all insects are easily identified alive. It is sometimes necessary to kill and preserve your specimens either to identify later or to form a reference collection for future identification. Immersion in 70% industrial methylated spirit will usually both kill and preserve soft-bodied insects such as aphids or larvae of moths, beetles and sawflies. Remember to label each vial clearly with collection date, site, collector's name and the insect's identity, when known. Hard-bodied insects, such as adult beetles or lepidopterans, may be killed in a closed jar or vial containing blotting or filter paper which has been moistened with a few drops of ethyl acetate. Alternatively, a killing jar in which leaves from the cherry laurel are used instead of ethyl acetate is more environmentally friendly and in keeping with the theme of this book! It is important to remember that the cyanide given off by the leaves is poisonous to humans as well as insects. The insect specimens may then be mounted on card or directly on a pin. Smaller insects may be card-mounted by gluing the feet and antennae to a small piece of thin card or transparent acetate sheet with water-soluble glue and mounting the card or acetate on a pin. Once again, make sure that all specimens are fully labelled with collection date, collection site, the name of the collector and the name of the species. If allowed to dry out after death, insects tend to become stiff and thus difficult to mount in a 'natural' position. They can be relaxed by exposing them for a few hours to a humid atmosphere, for example by keeping them in a closed vial with moist filter paper – but if kept humid for too long they may go mouldy.

Rearing insects in the laboratory

Insect rearing cages can easily be constructed from old ice-cream cartons. Punch small holes in the lids. Make sure that the holes are smaller than the insects, otherwise your specimens will almost certainly escape. If you suspect that insects are escaping through the holes, glue small pieces of muslin over them to prevent any further losses. Alternatively, just use the bases of the same cartons, covered with a piece of netting or muslin held in place with an elastic band. Either method allows easy access to the insects, but translucent muslin or netting has the extra advantage of allowing you to see inside the container without removing the top, which can sometimes be useful. Certain chocolate companies sell their products in transparent plastic containers with tight-fitting lids. These containers make ideal insect cages, although it is advisable to make small holes in the lids.

Other insect cages fit over the top of plant pots containing small branches or saplings. These cages are available from entomological suppliers or can be constructed from thick but flexible PVC sheeting, bent into a cylinder of the desired dimensions and glued with a suitable plastic

Fig. 18. A home-made insect cage covering a potted plant.

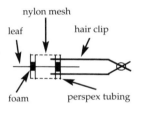

Fig. 19. A clip cage on a leaf.

Fig. 20. Ovary, showing ovarioles.

adhesive (fig. 18). The top is sealed with a muslin lid glued around the edge. A plastic fizzy drink bottle with the top and bottom cut off, and muslin glued across one end, will do just as well.

Clip cages are useful for working with small insects *in situ* on plants either out of doors or in the greenhouse or laboratory. A clip cage (fig.19) is constructed from a spring-loaded hair-clip, two short lengths of rigid perspex tubing and two rings of foam rubber. First cut two rings from the perspex tubing; they can be the same depth or one can be shallower than the other. Glue fine muslin onto the outer side of each ring and glue a ring of foam rubber round the inner rim of each ring. The pointed ends of the hair-clip are heated gently in a Bunsen burner flame until they are just glowing and then stuck into the perspex ring and held firmly until they set into position. The result is a neat efficient way of confining small insects such as aphids to individual leaves.

For handling small soft-bodied insects such as aphids and young caterpillars, one of the most useful aids is a fine camel hair paintbrush of size 00, 01 or 02. A pair of fine watchmakers forceps, although expensive, is well worth having, especially when picking up lepidopteran adults.

Care is necessary when introducing new food material to an insect culture. It is very easy to add unwanted insect species to the culture at the same time. For example, when rearing aphids, watch out for predators such as ladybird adults, larvae, or eggs, which can also be found on cherry trees, and also for small parasitic wasps. Check all foliage carefully before putting it in with insect specimens.

Ovariole number in aphids

As mentioned earlier in the book, it is sometimes very useful to be able to relate reproductive parameters of aphids to their host plants. Ovariole number varies in aphids, both within and between species and also within morphs. Although the prospect of dissecting an aphid to count its reproductive structures may seem daunting, it is actually relatively simple. The aphid can be gently anaesthetized using carbon dioxide from either a modified Sodastream apparatus (Killick-Kendrick, 1993) or one of the commercially available carbon dioxide wine-bottle openers. Place the aphid under a dissecting microscope, in a Petri dish with a small amount of water, and remove the entire reproductive system by gently pulling the segments at the tip of the abdomen with one pair of fine forceps, whilst holding the thorax with another. The number of ovarioles, the number of mature embryos (those with pigmented eyes) and the total number of embryos can then be estimated quite accurately (fig. 20).

Similarity indices

The Sorensen similarity coefficient is probably the easiest and most useful similarity index to use and has the form

$$Cs = 2j/(a+b)$$

where a and b are respectively the total number of species in each of two samples, and j is the number of species common (joint) to the two samples. Similarity indices can be used to see how similar the trees are to each other in terms of their insect fauna. Probably the easiest way to approach this is to construct a dendrogram.

This can be illustrated using the information in table 1 (p. 9). First, a similarity coefficient is calculated for all possible pairs of cherry species, and these coefficients are set out in a table. The two species showing the highest similarity coefficient in the table are named on a chart, with a vertical scale from 0 to 1, and they are joined at a level corresponding with their similarity coefficient. The next species to be entered is the one with the highest similarity to either of these, and so on. This process continues until all species have been entered (fig. 10, p. 15).

Studies of insect performance

The insects feeding on cherry trees can be used in two ways, either for experiments on the insects themselves, for example, examining the effect that host plant species or cultivar has on the growth and development of a particular insect species, or to examine the effect that the insects have on the trees.

Adult insects that feed on leaves or caterpillars can be reared on whole trees outside or on small trees in the greenhouse or laboratory. This is best achieved by placing around a branch a muslin sleeve securely tied around the branch at the trunk end and sealed with a flexi-tie or plant tie at the far end. Caterpillars can then be placed in the bags individually or in groups of ten or so, and allowed to feed on the leaves inside for a set time, say three weeks. The survivors are then collected from the bags and taken to the laboratory to be counted and weighed. Their number and size will give an index of the suitability of that tree species as a host plant. Studies of this kind should involve at least ten caterpillars per sleeve, but not more than 20 because at high densities competition for food may occur and insect viruses may become established.

Alternatively, the insects can be counted and measured *in situ* at regular intervals. These records can be used to compare rates of mortality and growth on different host plants.

If the work is based mainly in the laboratory, it is possible to explore the effects of leaf age on survival by comparing leaves that have just flushed with mature leaves or mature leaves with senescent leaves. Food quality may also affect adult insects such as chrysomelid beetles. Aphids are even easier to rear and study experimentally, especially if clip cages are used. It is fairly simple to manipulate plant seasonality if potted saplings and a greenhouse or cold store are available, and it may also be possible to examine the effect of temperature on the life-history traits of aphids. This is straightforward if controlled environment cabinets with both lighting and temperature controls are available.

Publishing your results

Writing up is perhaps one of the most important parts of a research project. A really thorough, critical investigation that has established new information of general interest may be worth publishing. Several journals may publish short articles on these kinds of investigations. Such journals include the *Entomologist's Monthly Magazine* or various local publications like *The Naturalist* (a journal which concentrates on work of relevance to Yorkshire). *The Journal of Biological Education* deals with articles of an educational nature including accounts of project work which may be a source of ideas for others. Examine recent numbers of these journals to see the sort of thing they publish, and then write a paper on similar lines, keeping it as short and concise as possible while still presenting enough information to establish the conclusions. Advice from an expert is helpful at this stage.

It is an unbreakable convention of scientific publication that results are reported with scrupulous honesty. It is therefore essential to keep detailed and accurate records throughout the investigation, and to distinguish in the write-up between certainty and probability, and between deduction and speculation.

It should be remembered that correlation between a habitat or microclimatic parameter and an aspect of invertebrate community structure may not necessarily reflect a direct causal relationship. The community structure may correlate well with leaf size, yet microclimate may be the crucial factor.

It will often be necessary to apply statistical techniques to test the significance of the findings. Books such as the *Open University Project Guide* (Chalmers & Parker, 1989) or *Statistics for Biology* (Bishop, 1966) will help, but it is an area where expert advice can contribute much to the planning, as well as the analysis, of the work.

Some useful addresses

Suppliers of entomological equipment
Watkins and Doncaster, P.O. Box 5, Cranbrook, Kent
TN18 5EZ
Worldwide Butterflies Ltd., Compton House, nr. Sherborne,
Dorset DT9 4QN

Gelatin capsules from:
Farillon Ltd., Bryant Avenue, Gidea Park, Romford RM3 0PJ

**Polyethylene foam for staging insects (and for lining
storage boxes) from:**
Wilford Polyformes, Greaves Way, Stanbridge Road,
Leighton Buzzard, Bedfordshire

Suppliers of entomological books
New and secondhand
E.W. Classey Ltd., P.O. Box 93, Faringdon, Oxfordshire
SN7 7DR
L. Christie, 129 Franciscan Road, Tooting, London
SW17 8DZ
Pendleside Books, 359 Wheatley Lane Road, Fence, Burnley
BB12 9QA
Wyseby House Books, Oxdrove, Burghclere, Newbury,
Berkshire RG15 9JS

New books
NHBS, 2–3 Wills Road, Totnes, Devon TQ9 5XN

Entomological societies
Amateur Entomologists' Society, 355 Hounslow Road,
Hanworth, Feltham, Middlesex TW13 5JH

British Entomological and Natural History Society, Dinton
Pastures Country Park, Davis Street, Hurst, Reading,
Berkshire RG10 0TH

Royal Entomological Society, 41 Queen's Gate, London
SW7 5HU

British Plant Gall Society, Honorary Secretary (Dr C.K.
Leach), 1 Palfreyman Lane, Oadby, Leicester LE2 4UR

Checklist of insects of cherry trees

This list gives the names of species known to feed on cherry in Britain. The names are in the conventional form with the author and date of the first description of the species. Unless stated otherwise, names follow the checklists published by the Royal Entomological Society (Kloet & Hincks, 1964, 1975, 1977, 1978), in which other synonyms are also given.

LEPIDOPTERA
(names after Emmet, 1991; see Kloet & Hincks (1972) for synonyms)

Nepticulidae
  Ectoedemia atricollis (Stainton, 1857)
  Stigmella prunetorum (Stainton, 1855)
Tischeriidae
  Tischeria gaunacella (Duponchel, 1843)
Lyonetiidae
  Lyonetia clerkella (Linnaeus, 1758)
Gracillariidae
  Phyllonorycter cerasicolella (Herrich-Schäffer, 1855)
  P. sorbi (Frey, 1855)
Yponomeutidae
  Argyresthia pruniella (Clerck, 1759)
  Scythropia crataegella (Linnaeus, 1767)
  Swammerdamia caesiella (Hübner, 1796)
  S. pyrella (Villers, 1789)
  Yponomeuta evonymella (Linnaeus, 1758)
  Y. padella (Linnaeus, 1758)
Coleophoridae
  Coleophora adjectella Herrich-Schäffer, 1861
  C. anatipenella (Hübner, 1796)
    = *bernoulliella* (Goeze, 1783), *sensu* Schnack, 1985
  C. coracipennella (Hübner, 1796)
  C. hemerobiella (Scopoli, 1763)
  C. prunifoliae (Doets, 1944)
  C. spinella (Schrank, 1802)
    = *cerasivorella* Packard, 1870
  C. trigeminella Fuchs, 1881
Tortricidae
  Acleris rhombana (Denis & Schiffermüller, 1775)
  A. notana (Donovan, 1806)
  A. umbrana (Hübner, 1796–99)
  Adoxophyes orana (Fischer v. R., 1834)
  Cydia funebrana (Treitschke, 1835)
  C. prunivorana (Ragonot, 1879)
  Enarmonia formosana (Scopoli, 1763)
  Epiblema trimaculana (Haworth, 1811)
  Epinotia signatana (Douglas, 1845)
  Hedya dimidioalba (Retzius, 1783)
    = *nubiferana* (Haworth, 1811)
  H. pruniana (Hübner, 1796–99)

Pammene germmana (Hübner, 1796–99)
P. rhediella (Clerck, 1759)
Pandemis cerasana (Hübner, 1786)
P. corylana (Fabricius, 1794)
Phtheochroa schreibersiana (Frölich, 1828)
Nymphalidae
Nymphalis polychloros (Linnaeus, 1758)
Geometridae
Abraxas grossulariata (Linnaeus, 1758)
Alsophila aescularia (Denis & Schiffermüller, 1775)
Lomographa bimaculata (Fabricius, 1775)
L. temerata (Denis & Schiffermüller, 1775)
Operophtera brumata (Linnaeus, 1758)
Triphosa dubitata (Linnaeus, 1758)
Notodontidae
Diloba caeruleocephala (Linnaeus, 1758)
Noctuidae
Allophyes oxyacanthae (Linnaeus, 1758)
Xestia baja (Denis & Schiffermüller, 1775)

DIPTERA

Cecidomyiidae
Aphidoletes aphidimyza (Rondani, 1847)
Asphondylia prunorum (Wachtl, 1888)
Dasineura tortrix (Löw, F., 1877)
Putoniella pruni (Kaltenbach, 1872)
Agromyzidae
Phytobia cerasiferae (Kangas, 1955)
Tephritidae
Rhagoletis cerasi (Linnaeus, 1758)

HYMENOPTERA

Pamphiliidae
Pamphilius sylvaticus (Linnaeus, 1758)
Neurotoma saltuum (Linnaeus, 1758)
Tenthredinidae
Caliroa cerasi (Linnaeus, 1758)
Pristiphora retusa (Thomson, 1871)

COLEOPTERA

Buprestidae
Anthaxia nitidula (Linnaeus, 1758)
Trachys minutus (Linnaeus, 1758)
Nitidulidae
Meligethes coracinus Sturm, 1845
Cerambycidae
Saperda scalaris (Linnaeus, 1758)
Chrysomelidae
Phytodecta pallida (Linnaeus, 1758)
Attelabidae
Rhynchites (Caenorhinus) aequatus (Linnaeus, 1767)
R. (C.) pauxillus Germar, 1824

R. *(Rhynchites) auratus* (Scopoli, 1763)
R. *(R.) bacchus* (Linnaeus, 1758)
R. *(R.) caeruleus* (Degeer, 1775)
Curculionidae
    Anthonomus bituberculatus Thomson, S.G., 1868
    A. humeralis (Panzer, 1795)
    Furcipus rectirostris (Linnaeus, 1758)
    Magdalis cerasi (Linnaeus, 1758)
    M. ruficornis (Linnaeus, 1758)
    Ramphus pulicarius (Herbst, 1795)
    Rhychaenus testaceus (Müller, O.F., 1776)
Scolytidae
    Scolytus mali (Bechstein, 1805)
    S. rugulosus (Müller, P.W.J., 1818)

HEMIPTERA

HETEROPTERA

Pentatomidae
    Pentatoma rufipes (Linnaeus, 1758)
Miridae
    Globiceps (Paraglobiceps) flavomaculatus (Fabricius, 1794)
    Lygocoris pabulinus (Linnaeus, 1761)
    Orthotylus marginalis Reuter, 1883
    Phytocoris dimidiatus Kirschbaum, 1856

HOMOPTERA

Cicadellidae
 (names follow Le Quesne & Payne (1981); see Kloet &
Hincks (1964) for synonyms)
    Aguriahana stellulata (Burmeister, 1841)
    Alebra wahlbergi (Boheman, 1845)
    Alnetoidia alneti (Dahlbom, 1850)
    Edwardsiana crataegi (Douglas, 1876)
    E. prunicola (Edwards, 1914)
    E. rosae (Linnaeus, 1758)
    Typhlocyba quercus (Fabricius, 1777)
    Zygina flammigera (Geoffroy, 1785)
Aphididae
    Brachycaudus cardui (Linnaeus, 1758)
    B. helichrysi (Kaltenbach, 1843)
    Hyalopterus pruni (Geoffroy, 1762)
    Myzus (Myzus) cerasi (Fabricius, 1775)
    Myzus (Nectarosiphon) persicae (Sulzer, 1776)
    Myzus (Prunomyzus) padellus Rog. & Hille, 1946
    Phorodon humuli (Schrank, 1801)
    Rhopalosiphum insertum (Walker, 1849)
    R. padi (Linnaeus, 1758)

Further reading

Finding books

Some of the books and journals listed here will be unavailable in local and school libraries. It is possible to make arrangements to see or borrow such works by seeking permission to visit the library of a local university or by asking your local public library to borrow the work (or a photocopy of it) for you via the British Library Lending Division. This may take several weeks. It is important to present your librarian with a reference that is correct in every detail. References are acceptable in the form given here, namely the author's name and date of publication, followed by (for a book) the title and publisher or (for a journal article) the title of the article, the full journal title, the volume number, and the first and last pages of the article.

Bakker, K. (1961) An analysis of factors which determine success in competition for food among larvae of *Drosophila melanogaster*. *Archives Neerlandaises Zoologie* **14**: 200-281.

Baldwin, I.T. & Schultz, J.C. (1983) Rapid changes in tree leaf chemistry induced by damage: evidence for communication between plants. *Science* **221**: 277-279.

Barnes, H.F. (1948) *Gall Midges of Economic Importance. Volume 3 Gall Midges of Fruit.* London: Crosby Lockwood.

Barnes, H.F. (1951) *Gall Midges of Economic Importance. Volume 5 Gall Midges of Trees.* London: Crosby Lockwood.

Benson, R.B. (1951) Hymenoptera. 2. Symphyta Section (a). *Handbooks for the Identification of British Insects* **6**(2a). Royal Entomological Society of London.

Benson, R.B. (1952) Hymenoptera. 2. Symphyta Section (b). *Handbooks for the Identification of British Insects* **6**(2b). London: Royal Entomological Society of London.

Benson, R.B. (1958) Hymenoptera. 2. Symphyta Section (c). *Handbooks for the Identification of British Insects* **6**(2c). London: Royal Entomological Society of London.

Birks, H.J.B. (1980) British trees and insects: a test of the time hypothesis over the last 13,000 years. *American Naturalist* **115**: 600–605.

Birks, H.J.B. (1989) Holocene isochrone maps and patterns of tree-spreading in the British Isles. *Journal of Biogeography* **16**: 503–540.

Birks, H.J.B., Deacon, J. & Peglar, S.M. (1975) Pollen maps for the British Isles 5000 years ago. *Proceedings of the Royal Society of London B* **189**: 87–105.

Bishop, O.N. (1966) *Statistics for Biology.* London: Longman.

Blackman, R.L., (1974) *Aphids.* London: Ginn & Co. Ltd.

Blackman, R.L. & Eastop, V.F. (1994) *Aphids of the World's Trees.* Wallingford: CABI.

Bradley, J.D., Tremewan, W.G. & Smith, A. (1973) *British Tortricoid Moths. Cochylidae and Tortricidae: Tortricinae.* London: Ray Society.

Bradley, J.D., Tremewan, W.G. & Smith, A. (1979) *British Tortricoid Moths. Tortricidae: Olethreutinae.* London: Ray Society.

Brooks, M. (1991) *A Complete Guide to British Moths (Macrolepidoptera).* London: Jonathan Cape.

Buckler, W. (1886–1901) *Larvae of British Butterflies and Moths, Volumes 1–9.* London: Ray Society.

Cameron, P. (1882–1892) *A Monograph of the British Phytophagous Hymenoptera Volumes 1–4.* London: Ray Society.

Carter, D.J. & Hargreaves, B. (1986) *A Field Guide to the Caterpillars of Butterflies and Moths in Britain and Europe.* London: Collins.

Chalmers, N. & Parker, P. (1989) *The OU Project Guide.* (2nd edn.) Field Studies Occasional Publications No. 9. Shrewsbury: Field Studies Council.

Claridge, M.F. & Evans, H.F. (1990) Species-area relationships: relevance to pest problems of British trees? In *Population Dynamics of Forest Insects* (ed. A.D. Watt, S.R. Leather, M.D. Hunter & N.A.C. Kidd), pp 59–69. Andover: Intercept Press.

Colyer, C.N. & Hammond, C.O. (1968, 2nd edn.) *Flies of the British Isles*. London: Frederick Warne.

Connell, J.H. (1980) Diversity and the coevolution of competitors, or the ghost of competition past. *Oikos*, **35**: 131–138.

Cox, C.B., Healey, I.N. & Moore, P.D. (1976) *Biogeography*. (2nd edn.) Oxford: Blackwell Scientific Publications.

Darlington, P.J. (1957) *Zoogeography: The Geographical Distribution of Animals*. New York: John Wiley & Sons.

Dixon, A.F.G. (1985) *Aphid Ecology*. Glasgow: Blackie.

Duffy, E.J. (1952) Coleoptera, Cerambycidae. *Handbooks for the Identification of British Insects* **5**(12). London: Royal Entomological Society of London.

Duffy, E.J. (1953) Coleoptera, Scolytidae and Platypodidae. *Handbooks for the Identification of British Insects* **5**(15). London: Royal Entomological Society of London.

Edwards, P.J., Wratten, S.D. & Parker, E.A. (1992) The ecological significance of rapid wound-induced changes in plants: insect grazing and plant competition. *Oecologia* **91**: 266–272.

Emmet, A.M. (1991) Chart showing the life history and habits of the British Lepidoptera. In *The Moths and Butterflies of Great Britain and Ireland. Volume 7(2)* (eds. Emmet, A.M. & Heath, J.) pp. 61–303. Colchester: Harley Books.

Emmet, A.M. & Heath, J. (1989) *The Moths and Butterflies of Great Britain and Ireland. Volume 7 part 1, Hesperiidae–Nymphalidae*. Colchester: Harley Books.

Emmet, A.M. & Heath, J. (1991) *The Moths and Butterflies of Great Britain and Ireland. Volume 7, part 2, Lasiocampidae–Thyatiridae*. Colchester: Harley Books.

Emmet, A.M. & Heath, J. (1996) *The Moths and Butterflies of Great Britain and Ireland. Volume 3, Yponomeutidae–Elachistidae*. Colchester: Harley Books.

Emmet, A.M., Langmaid, J.R., Bland, K.P., Corley, M.F.V. & Razowski, J. (1996) Coleophoridae. In *The Moths and Butterflies of Great Britain and Ireland, Volume 3* (ed. A.M. Emmet) pp. 126–338. Colchester: Harley Books.

Fowler, S.V. & Lawton, J.H. (1982) The effects of host-plant distribution and local abundance on the species richness of agromyzid flies attacking British umbellifers. *Ecological Entomology* **7**: 257–265.

Fowler, S.V. & Lawton, J.H. (1985) Rapidly induced defences and talking trees: the devil's advocate position. *American Naturalist* **126**: 181–195.

Goater, B. (1986) *British Pyralid Moths, A Guide to their Identification*. Colchester: Harley Books.

Godwin, H. (1975) *The History of the British Flora: a Factual Basis for Phytogeography*. (2nd edn.) Cambridge: Cambridge University Press.

Gorman, M. (1979) *Island Ecology*. London: Chapman & Hall.

Haggett, G.M. (1981) *Larvae of the British Lepidoptera not Figured by Buckler*. London: British Entomological and Natural History Society.

Haukioja, E. (1991) Cyclic fluctuations in density – interactions between a defoliator and its host tree. *Acta Oecologica* **12**: 77-88.

Heath, J. (1976) *The Moths and Butterflies of Great Britain and Ireland. Volume 1, Micropterigidae–Heliozelidae*. Colchester: Harley Books.

Heath, J. & Emmet, A.M. (1979) *The Moths and Butterflies of Great Britain and Ireland. Volume 9, Sphingidae–Noctuidae (part 1)*. Colchester: Harley Books.

Heath, J. & Emmet, A.M. (1979) *The Moths and Butterflies of Great Britain and Ireland. Volume 10, Sphingidae–Noctuidae (part 2)*. Colchester: Harley Books.

Heath, J. & Emmet, A.M. (1985) *The Moths and Butterflies of Great Britain and Ireland. Volume 2, Cossidae–Heliodinidae*. Colchester: Harley Books.

Heie, O. (1986) *The Aphidoidea (Hemiptera) of Fennoscandia and Denmark III*. Fauna Entomologica Scandinavica 17.

Heie, O. (1992) *The Aphidoidea (Hemiptera) of Fennoscandia and Denmark IV*. Fauna Entomologica Scandinavica 25.

Heie, O. (1994) *The Aphidoidea (Hemiptera) of Fennoscandia and Denmark V.* Fauna Entomologica Scandinavica 28.

Henderson (1979) *Dictionary of Biological Terms.* London: Penguin.

Hodge, P.J. & Jones, R.A. (1995) *New British Beetles. Species not in Joy's Practical Handbook.* London: British Entomological and Natural History Society.

Hutchinson, G.E. (1944) Limnological studies in Connecticut: critical examination of the supposed relationship between phytoplanktonic periodicity and chemical changes in lake waters. *Ecology* **25**: 3–26.

Jones, C.G. & Lawton, J.H. (1991) Plant chemistry and insect species richness of British Umbellifers. *Journal of Animal Ecology* **60**: 767–777.

Joy, N.H. (1932) *A Practical Handbook of British Beetles, Volumes 1 &2.* London: Witherby.

Karban, R. & Baldwin, I. T. (1997) *Induced Responses to Herbivory.* Chicago: University of Chicago Press.

Kennedy, C.E.J. & Southwood, T.R.E. (1984) The number of species of insects associated with British trees: a re-analysis. *Journal of Animal Ecology* **53**: 455-478.

Killick-Kendrick, R. (1993) 'Sodastream' CO_2 dispenser for killing insects. *Antenna* **17**: 115.

Kirk, W. D. J. (1992) *Insects on Cabbages and Oilseed Rape.* Naturalists' Handbooks 18. Slough: The Richmond Publishing Co. Ltd.

Kirk-Spriggs, A.H. (1996) Pollen Beetles. Coleoptera: Kateretidae and Nitidulidae: Meligethinae. *Handbooks for the Identification of British Insects* 5(6a). London: Royal Entomological Society.

Kloet, G.S. & Hincks, W.D. (1964) A Check List of British Insects Part 1. Small Orders and Hemiptera. *Handbooks for the Identification of British Insects* **11**(1). London: Royal Entomological Society of London.

Kloet, G.S. & Hincks, W.D. (1972) A Check List of British Insects Part 2. Lepidoptera. *Handbooks for the Identification of British Insects* **11**(2). London: Royal Entomological Society of London.

Kloet, G.S. & Hincks, W.D. (1976) A Check List of British Insects Part 5. Diptera and Siphonoptera. *Handbooks for the Identification of British Insects* **11**(5). London: Royal Entomological Society of London

Kloet, G.S. & Hincks, W.D. (1977) A Check List of British Insects Part 3. Coleoptera and Strepsiptera. *Handbooks for the Identification of British Insects* **11**(3). London: Royal Entomological Society of London.

Kloet, G.S. & Hincks, W.D. (1978) A Check List of British Insects Part 4. Hymenoptera. *Handbooks for the Identification of British Insects* **11**(4). London: Royal Entomological Society of London.

Lawton, J.H. (1983) Plant architecture and the diversity of phytophagous insects. *Annual Review of Entomology* **28**: 23–39.

Lawton, J.H. (1984) Non-competitive populations, non-convergent communities and vacant niches: the herbivores on bracken. In *Ecological Communities: Conceptual Issues and the Evidence* (eds. D.R. Strong, D. Simberloff, L.G. Abele & A.B. Thistle), pp 67–100. Princeton, USA: Princeton University Press.

Lawton, J.H., Cornell, H., Dritschilo, W. & Hendrix, S.D. (1981) Species as islands: comments on a paper by Kuris *et al. American Naturalist* **117**: 623–627.

Lawton, J.H. & Price, P.W. (1979) Species richness of parasites on hosts: Agromyzid flies on the British Umbelliferae. *Journal of Animal Ecology* **48**: 619–637.

Lawton, J.H. & Schroder, D. (1977) Effects of plant type, size of geographical range and taxonomic isolation on number of insect species associated with British plants. *Nature* **265**: 137–140.

Leather, S.R. (1985) Does the bird cherry have its 'fair share' of insect pests? An appraisal of the species–area relationships of the phytophagous insects associated with British *Prunus* species. *Ecological Entomology* **10**: 43–56.

Leather, S.R. (1986) Insect species richness of the British Rosaceae: the importance of host range, plant architecture, age of establishment, taxonomic isolation and species-area relationships. *Journal of Animal Ecology* **55**: 841-860.

Leather, S.R. (1987) Generation specific trends in aphid life history parameters. *Journal of Applied Entomology* **104**: 278–284.

Leather, S.R. (1988) Consumers and plant fitness: coevolution or competition? *Oikos* **53**: 285–288.

Leather, S.R. (1990*a*) The analysis of species-area relationships with particular reference to macrolepidoptera on Rosaceae: how important is data set quality? *The Entomologist* **109**: 8–16.

Leather, S.R. (1990*b*) The role of host quality, natural enemies, competition and weather in the regulation of autumn and winter populations of the bird cherry aphid. In *Population Dynamics of Forest Insects* (eds. A. D. Watt, S. R. Leather, M. D. Hunter & N. A. C. Kidd), pp 35–44. Andover: Intercept Press.

Leather, S.R. (1992) Aspects of aphid overwintering (Homoptera: Aphidinea: Aphididae) *Entomologia Generalis* **17**: 101-113.

Leather, S.R. (1996). Biological Flora of the British Isles. *Prunus padus* L. *Journal of Ecology* **84**: 125–132.

Leather, S.R .& & Lehti, J.P. (1982) Abundance and distribution of *Yponomeuta evonymellus* (Lepidoptera, Yponomeutidae) in Finland during 1981. *Notulae Entomologica* **62**: 93–96.

Leather, S.R., Wellings, P.W. & Dixon, A.F.G. (1983) Habitat quality and the reproductive strategies of the migratory morphs of the bird cherry-oat aphid, *Rhopalosiphum padi* (L.), colonizing secondary host plants. *Oecologia* **59**: 302-306.

Le Quesne, W.J. & Payne, K.R. (1981) Cicadellidae (Typhlocybinae) with a check list of the British Auchenorhyncha (Hemiptera, Homoptera). *Handbooks for the Identification of British Insects* 2(2c). London: Royal Entomological Society of London.

Levey, B. (1977) Coleoptera, Buprestidae, *Handbooks for the Identification of British Insects* 5(1b). London: Royal Entomological Society of London.

Mitchell, A. (1974) *A Field Guide to the Trees of Britain and Northern Europe*. London: Collins.

Moran, V.C. (1980) Interactions between phytophagous insects and their *Opuntia* hosts. *Ecological Entomology* **5**: 153–164.

Morris, W.G. (1991) *Weevils*. Naturalists' Handbooks 16. Slough: The Richmond Publishing Co. Ltd.

Morse, D.R., Lawton, J.H., Dodson, M.M. & Williamson, M.H. (1985) Fractal dimension of vegetation and the distribution of arthropod body lengths. *Nature* **314**: 731–733.

Neuvonen, S., Haukioja, E. & Molarius, A. (1987) Delayed inducible resistance against a leaf chewing insect in four deciduous tree species. *Oecologia* **74**: 363–369.

Opler, P.A. (1974) Oaks as evolutionary islands for leaf-mining insects. *American Scientist* **62**: 67–73.

Perring, F.H. & Walters, S.M. (1962) *Atlas of the British Flora*. London: Nelson.

Price, P.W. (1975) *Insect Ecology*. New York: John Wiley & Sons.

Read, R.W.J. (1981) *Furcipus rectirostris* (L.) (Coleoptera: Curculionidae) new to Britain. *Entomologist's Gazette* **32**: 51–58.

Rhoades, D.F. (1983) Responses of alder and willow to attack by tent caterpillars and webworms: evidence for pheromonal sensitivity of willows. *American Chemical Society Symposium Series* **208**: 55–68.

Root, R.B. (1967) The niche exploitation pattern of the blue-gray gnatcatcher. *Ecological Monographs* **37**: 317–350.

Rotheray, G.E. (1993) *Colour Guide to Hoverfly Larvae (Diptera, Syrphidae) in Britain and Europe*. Dipterist's Digest **9**.

Skinner, B. (1984) *Colour Identification Guide to Moths of the British Isles*. Middlesex: Viking.

Skinner. G.J. & Allen, G.W. (1996) *Ants*. Naturalists' Handbooks 24. Slough: The Richmond Publishing Co. Ltd.

Smith, K.G.V. (1989) An Introduction to the Immature Stages of British Flies. *Handbooks for the Identification of British Insects* **10**(4). London: Royal Entomological Society of London.

Southwood, T.R.E. (1961) The number of species of insects associated with various trees. *Journal of Animal Ecology* **30**: 1–8.

Southwood, T.R.E. & Leston, D. (1959) *Land and Water Bugs of the British Isles.* London: Frederick Warne.

Spencer, K.A. (1972) Diptera, Agromyzidae. *Handbooks for the Identification of British Insects* **10**(5g). London: Royal Entomological Society of London.

Stace, C. (1997) *New Flora of the British Isles.* (2nd edn.) Cambridge: Cambridge University Press.

Stokoe, W.J. & Stovin, G.H.T. (1948) *The Caterpillars of British Moths, Volumes 1 & 2.* London: Frederick Warne.

Strong, D.R. & Levin, D.A. (1979) Species richness of plant parasites and growth form of their hosts. *American Naturalist* **114**: 1–22.

Strong, D.R., Lawton, J.H. & Southwood, T.R.E. (1984) *Insects on Plants.* Oxford: Blackwell Scientific Publications.

Stroyan, H.L.G. (1984) Aphids – Pterocommatinae and Aphidinae (Aphidini), Homoptera, Aphididae. *Handbooks for the Identification of British Insects* 2(6). London: Royal Entomological Society of London.

Stubbs, A.E. & Falk, S.J. (1983) *British Hoverflies. An Illustrated Identification Guide.* London: British Entomological and Natural History Society.

Thomas, J. & Lewington, R. (1991) *The Butterflies of Britain and Ireland.* London: Dorling Kindersley.

Unwin, D.W. (1981) A Key to the Families of British Diptera. Field Studies Council Publication no. 143. *Field Studies* **5**(3): 513–553. (An AIDGAP key)

Unwin, D.W. (1988) A Key to the Families of British Coleoptera (and Strepsiptera). Field Studies Council publication no. 166. *Field Studies* **6**(4): 149–197. (An AIDGAP key)

Van Emden, H.F, (1990) Limitations on insect herbivore abundance in natural vegetation. In *Pests, Pathogens and Plant Communities* (eds. J.J. Burdon & S.R. Leather), pp 15–30. Oxford: Blackwell Scientific Publications.

White, I.M. (1988) Tephritid Flies. Diptera: Tephritidae. *Handbooks for the Identification of British Insects* 10(5a). London: Royal Entomological Society of London.

Whitham, T.G. (1978) Habitat selection by *Pemphigus* aphids in response to resource limitation and competition. *Ecology* **59**: 1164–1176.

Wright, A. (1990) British sawflies (Hymenoptera: Symphyta). A key to the adults of genera occurring in Britain. *Field Studies* **7**(3): 531–593. (An AIDGAP key)

Whitham, T.G. (1986) Costs and benefits of territoriality: behavioral and reproductive release by competing aphids. *Ecology* **67**: 139–147.

INDEX

In addition to the references below, there is a checklist of insect species that live on cherry trees on pp. 72–74